Proving It Her Way

David E. Rowe • Mechthild Koreuber

Proving It Her Way

Emmy Noether, a Life in Mathematics

 Springer

David E. Rowe
Mathematics Institute
Mainz University
Mainz, Germany

Mechthild Koreuber
Chief gender equality officer
Freie Universität Berlin
Berlin, Germany

ISBN 978-3-030-62810-9   ISBN 978-3-030-62811-6   (eBook)
https://doi.org/10.1007/978-3-030-62811-6

Mathematics Subject Classification (2020): 01A60, 01A70, 01A72, 01A73

Cover illustration: Emmy Noether, ca. 1920. SUB Göttingen, Sammlung Voit, no. 4.

This Springer imprint is published by the registered company Springer Nature Switzerland AG
The registered company address is: Gewerbestrasse 11, 6330 Cham, Switzerland

Contents

Preface

Producing a book normally involves a long, arduous process that some authors feel compelled to describe in its preface. We have no such inclination, but even if that were the case, the truth is that this book came about as an afterthought. It arose out of our longstanding fascination with the career of Emmy Noether, but also from recent circumstances and events that made this project too tempting to resist. In 2019, the second author organized a major conference at the Freie Universität Berlin in cooperation with the Berlin Mathematics Research Center MATH+ and the Max Planck Institute for the History of Science to commemorate the hundredth anniversary of Noether's Habilitation in Göttingen.[1] This event, which took place on June 4, 1919, was not only a major milestone for Noether herself but also for women longing to pursue academic careers in Germany.

As one of the highlights of the conference, the ensemble *portraittheater Vienna* presented the premiere performance of their play, "Mathematische Spaziergänge mit Emmy Noether" [Schüddekopf/Zieher 2019]. Both of us, as historians of mathematics, had been involved in its production, and we were delighted by the result. So the idea of adapting the script for an English-speaking audience occurred to us right away. It must be said, though, that neither of us had any idea how Sandra Schüddekopf and Anita Zieher would manage to stage a play about a mathematician, one whose work even her peers found to be highly abstract. Nevertheless, they found a very elegant way to finesse that problem, and in a manner that would have appealed to Emmy Noether, whose personality shines through despite the handicap that most of her audience has absolutely no idea what she's really talking about.

It is worth mentioning that most of those present at the opening that evening were *not* mathematicians, because this and subsequent performances have shown that nearly anyone can enjoy this play. It was written, in fact, with the idea of introducing Emmy Noether to audiences that lack all the necessary prerequisites for truly understanding her, save one: curiosity. Viewers need not know anything about higher mathematics in order to grasp why Noether was a uniquely dynamic figure, a woman whose life story deserves to be known beyond the small sphere of those who happen to have heard about commutative rings. She loved to walk and talk and swim, while all the time living out her great passion for mathematics. Most who knew her from Göttingen, where she became a fixture of its famous mathematical community, thought of Emmy Noether as a simple woman who thoroughly enjoyed the simple things in her life, whether doing mathematics or eating pudding with her students and friends.[2] This more visible, light-hearted demeanor was, however, only one side of Emmy Noether's personality, and she

---

[1] The interdisciplinary character of the conference – which brought together mathematicians, physicists, historians of science, gender researchers, and cultural historians – is reflected in the forthcoming conference volume [Koreuber 2021].

[2] The latter activity, a major motif in the play, is described in [Dick 1970/1981, 1981: 49]: "The pudding always tasted the same – delicious."

had to endure many sorrows in her life. Barely had she arrived in Göttingen in the spring of 1915, when she received the news of her mother's death. This was only the first of four trips she would make back to Erlangen to mourn the passing of a family member: her father and also two of her three younger brothers. On such occasions, her calm dignity greatly impressed those who witnessed how she carried herself. Emmy Noether was by no means a one-dimensional type who lived for mathematics and nothing else, and Anita Zieher (Fig. 1.1) truly brings her full personality back to life on stage, now in the new adaptation of the original play, "Diving into Math with Emmy Noether" [Schüddekopf/Zieher 2020].[3]

Our book is thus, in the first instance, a companion to this version of the play. In both cases, the aim is to provide a window on the life of an extraordinarily creative mathematician, taking due account that her mathematical world is only truly accessible to those familiar with higher mathematics. Naturally, we have made many allusions to that world which will resonate for readers acquainted with modern algebra and abstract mathematical conceptions, but we have refrained from offering any details or explanations. So the present book remains, so to speak, on the surface by offering only vague hints of Emmy Noether's "real world," a realm open only to those who have "satisfied the prerequisites." For that audience – those seeking a more comprehensive picture of Noether's work and its central place in the mathematical world of her time – we recommend the volume *Emmy Noether – Mathematician Extraordinaire* [Rowe 2020b], written for readers who, let us say, have a vague recollection of Noetherian rings. Even though these two books are intended for quite different audiences, the goal in both cases is the same: to illuminate Noether's life in the context of her times by conveying a full-blooded picture of her role in shaping the mathematical activity of her day and, as it happened, well beyond.

For many people – including some with otherwise broad cultural interests – the very idea that doing mathematics might be a truly creative intellectual activity sounds a bit strange, if not wholly absurd. Yet for Emil Artin, one of Emmy Noether's close friends and to some extent a kindred spirit, mathematics was an art, not a science.[4] This, to be sure, has long been a contested position; indeed, it was a favorite topic of friendly disputes between Emmy and her brother Fritz, a leading applied mathematician who worked on problems like modeling turbulence in dynamical systems. Their father, Max Noether, was a different type of mathematician still, as will be seen in Chapter 3, when we turn to their years together in Erlangen. Mathematical talent ran through the Noether family, which led the Göttingen number-theorist Edmund Landau to liken Emmy Noether's kinfolk with a coordinate system in which she occupied the origin [Dick 1970/1981, 1981: 95].

In a book like this one, it would be pointless to attempt anything like a theoretical discussion of various types of mathematical creativity. Instead, we

---

[3] For information or to book a performance, contact office@portraittheater.net.

[4] On the broader context of earlier debates over the status of mathematics as art or science, see [Rowe 2018a, 401–411].

simply want to claim that the label mathematician, when applied to elites like the Noethers, fails to convey the vast range of intellectual activity covered by the term. In the case of Emmy Noether, this is a crucially important point; her way of thinking about mathematics using abstract concepts, rather than concrete objects, was by no means new. She became, however, the foremost exponent of this approach to mathematical theorizing, which she promoted in a radical manner, a style quite unlike that of any other contemporary figure. She and Artin also both believed that all truly deep mathematical truths must be beautiful. One can only begin to understand what that means, of course, by delving deeply into the mathematical world they shared, as many famous figures who came after them did. We, on the other hand, will be satisfied if this striking and, surely for many people, surprising aspect – mathematical insight as a special type of aesthetic experience – is at least dimly reflected in our account of Noether's visionary work. Or, to put this somewhat more ambitiously, we hope to show that people like her, who engage in mathematical research at the highest levels, can and should be considered as artists of a special kind. Still, we can only illumine such creative work indirectly, through the framework of its more mundane manifestations.[5] We nevertheless hope that in this way the mysterious archipelago of mathematical theories and ideas – some dubbed "abstract nonsense" by mathematicians themselves – will appear more transparent, less strange, and above all more clearly rooted in the larger realm of human intellectual endeavors.[6]

Parallels between our book and the play abound throughout, so we begin by highlighting information directly relevant to what unfolds onstage. Thus, Chapter 1 offers some skeletal facts concerning Emmy Noether's life along with brief information about the four other mathematicians who appear in cameo roles. Quite a few other people are mentioned on stage as well – most of them familiar figures in the world of mathematics – so this chapter closes with a glossary identifying several others whose lives intersected with Noether's. All of them surface in various places in the chapters that follow. Chapter 2 then moves beyond the bare bones of Emmy Noether's life by providing a rounded portrait without undue attention to details. With these first two chapters in place as background information, we invite readers to dive down deeper in those that then follow. Although arranged chronologically, they also focus on important themes and historical events that shaped different phases of Noether's life.

Emmy Noether is justly famous as the "mother of modern algebra," but it is important to understand what she meant by "doing algebra." Her vision of its role in mathematics did not seek to erect clear disciplinary boundaries setting algebraic investigations off from those in other fields. On the contrary, her work was closely tied to a larger trend that aimed to algebraicize other fields, from complex functions and number theory to topology, i.e. major parts of all mathematical

---

[5]In this respect, we follow the lead of Carol Parikh, author of *The Unreal Life of Oscar Zariski* [Parikh 1991], his real life being, of course, his mathematical works.

[6]The same motivation is central to the earlier study [Koreuber 2015], which approaches Emmy Noether's unique influence from the perspective of philosophy of science.

knowledge. In this respect, she clearly identified with words she once cited from Leopold Kronecker's 1861 inaugural address, when he was inducted into the Berlin Academy: "algebra is not actually a discipline in itself but rather the foundation and tool of all mathematics."[7] Not that many of Noether's contemporaries shared this view; far from it. Nor should we imagine that Noether meant this literally; she was well aware of the vast fields in analysis and applied mathematics that lie well beyond the realm of even her all-embracing view of algebraic research. Still, she was clearly the leading spokesperson of her generation for a position that many of her contemporaries found extreme.

One of Noether's closest collaborators, Helmut Hasse, clearly recognized the import of Noether's message, but he also sensed the need to spread the word. In a lecture on "The modern algebraic Method," he said:

> The aim of my talk is to promote modern algebra among non-specialists instead of preaching to the choir. It is not my intention to lure anyone from his field of specialization to become an algebraist. I see my task, rather, as laying the groundwork for a favorable understanding of modern algebra, helping to establish its methods – insofar as they are of general importance, and integrating these methods into the common knowledge of contemporary mathematicians. [Hasse 1930, 22]

Noether's former Göttingen colleague, Hermann Weyl, on the other hand, had deep misgivings, though not so much with regard to abstract algebra per se. What concerned Weyl was the general trend toward abstraction in mathematical research. In his famous memorial address [Weyl 1935], delivered at Bryn Mawr College on April 26, 1935, twelve days after Noether's sudden death, Weyl alluded directly to their differences with regard to the fertility of abstraction. A number of other things Weyl said on that occasion have often been discussed in the by now quite vast literature on Emmy Noether, though many of his most insightful remarks have been overlooked or forgotten. Several commentators have derided Weyl for his condescending remarks about her lack of sophistication and femininity. Her friend Pavel Alexandrov repeated his remark that "no one could contend that the Graces had stood by her cradle," nodding in assent before he immediately went on to underscore what really mattered. Fräulein Noether, as all who knew her well could attest, "loved people, science, life with all the warmth, all the joy, all the selflessness and all the tenderness of which a deeply feeling heart – and a woman's heart – was capable" [Alexandroff 1935, 11].

Alexandrov's tribute to Emmy Noether took place in Moscow on a very different occasion, some five months after her death. He spoke (in Russian) as President of the Moscow Mathematical Society on 5 September 1935, during a week-long international conference on topology. One of the mathematicians in the audience, as acknowledged by the speaker, was Fritz Noether, Emmy's brother.

---

[7][Noether 1932c]; the relevance of this citation is discussed in [Koreuber 2015, 225] as well as in [Merzbach 1983, 161].

Soon after she arrived at Bryn Mawr, Emmy Noether had undertaken efforts to find a position for him in the United States, but without success (Section 9.3). He and his two sons, Hermann and Gottfried Noether, then emigrated from Breslau to the Soviet Union, where Fritz took up work at the Mathematical Institute in Tomsk.

When Alexandrov spoke, he began by noting Weyl's address as well as B.L. van der Waerden's obituary article [van der Waerden 1935] in order to say that his remarks would have a somewhat different character. "I would like to evoke for you as accurate an image as possible of the deceased, as a mathematician, as the head of a large scientific school, as a brilliant, original, fascinating personality" [Alexandroff 1935, 2]. Alexandrov spoke movingly about the woman he and his friend Pavel Urysohn first met in Göttingen in 1923. Her school then had only just begun and consisted of a mere handful of German students, but its international character would soon thereafter unfold. A major breakthrough came the very next year with the arrival of B.L. van der Waerden from Amsterdam. Alexandrov called him "one of the brightest young mathematical talents of Europe" and credited him with mastering Noether's theories, while adding significant results of his own, and "more than anyone else, [helping] to make her ideas widely known" [Alexandroff 1935, 5]. Three years later, van der Waerden taught a highly successful course on ideal theory in Göttingen, which did much to spread awareness of Noether's work. Some of Alexandrov's most vivid memories of Emmy came from the winter semester of 1928–1929 when they were together in Moscow. He spoke of her keen interest in the Soviet experiment and firm intention to visit again, but this was not to be.

As an intimate friend with longstanding ties to the Göttingen community during the 1920s, Alexandrov left a striking portrait of Emmy Noether as he knew her. What he, of course, did not pretend to know concerned her earlier life, about which rather little, in fact, is known. Hermann Weyl fully recognized the importance of her family background, but he noted at the outset of his memorial address that he could say little about Emmy's biography. Nor had he any chance to learn more about her life now that he, like her, was separated by an ocean from their homeland. Nevertheless, he tried to paint a picture of life in Erlangen as he imagined it to have been. He offered quite vivid portraits of the two contrasting personalities who dominated mathematical life in Erlangen: Max Noether and Paul Gordan, the first Emmy's father, the second her *Doktorvater*. He even tried to convey a sense of how their generation's sense of solidity surely "prevailed in the Noether home [leading to] a particularly warm and companionable family life.". This type of atmosphere, with "its comfort and bourgeois peacefulness" was now gone forever [Weyl 1935, 429].

Weyl clearly identified the Noether family, but Emmy in particular, as "good Germans" in a generic sense. Perhaps Weyl also felt unnerved by Emmy's apparent inability to grasp evil in the world. She had lived her whole life as a fully integrated German Jew, which meant of course that anti-Semitism was no stranger to her, but when the barbarians came to power and threatened to sweep away everything she loved, she reacted not only with restraint but with an almost super-human equanimity. Those lonely months during the spring of 1933 – the time when they were last together in Göttingen – no doubt profoundly shaped Weyl's view of her, yet his opinion seems to have wavered between two extremes: Emmy was either a tower of moral strength or she was simply naive.

Although Hermann Weyl knew Wilhelmian Germany exceedingly well, he may never have met Max Noether and he probably only got to know his daughter in 1927, when he spent a semester as guest professor in Göttingen. In attempting to explore Emmy Noether's social and intellectual background in this book, we have based our interpretation primarily on contemporary documentary sources, few of which, of course, were available to Weyl in 1935. We have also tried to avoid stereotypic themes found in much of the secondary literature. Many standard studies of women in the history of mathematics have chosen to follow Weyl's lead by comparing Noether with the internationally renowned Russian mathematician Sofia Kovalevskaya, who appears in several places in this book as well. These comparisons, to be sure, rarely have anything to do with serious interest in what these women accomplished as mathematicians. Very often, they are coupled together as two trailblazers in a field then totally dominated by men, even though neither really saw herself in such a role. Talk of glass ceilings, after all, was yet to come, whereas gender roles in that era were exceedingly constrained. Earlier commentators usually could not get past the notion that a "lady mathematician" was a freak of nature, a view clearly supported by the scarce number of these creatures then walking the earth. How that has changed! Contemporary opinions of Emmy Noether – and these were quite mixed – clearly have considerable importance for understanding the context in which she lived. Even more important – especially for our undertaking – are those sources that tell us how she thought about herself and the world around her, and especially how she expressed those thoughts. Others often compared her as a mathematician with Richard Dedekind (no one writing about her mathematics would have imagined a comparison with Sofia Kovalevskaya), but she quite rightly said about herself "I always went my own way" (see the opening of Chapter 2).

We have also departed from Weyl's interpretation in another important respect. Most standard accounts of Emmy Noether's intellectual development have followed his tripartite division of her career:[8] (1) the period as a post-doc, 1907–1919, followed by (2) her work on the general theory of ideals, 1920–1926, and then (3) her contributions to non-commutative algebras, their representation theory, and applications to commutative number fields, 1927–1935 [Weyl 1935, 439]. This periodization is certainly apt and useful to a point, but it also can easily lead to quite misleading impressions. Those who have adopted it have tended to underplay the significance of the first period, while overlooking some of the threads that ran through all three phases of Noether's career.

---

[8] Weyl's obituary of Hilbert was somewhat similar; there he discerned that the master's work fell into five periods [Weyl 1944, 4: 135].

Emmy Noether was nearly forty years old when she began publishing the papers on modern algebra that made her famous. By the mid-1920s, she had become the leader of an international school that would soon thereafter exert a deep and lasting influence on mathematical research. All her most familiar and significant work was thus undertaken during the latter two periods, when many of her ideas and findings quickly propagated through the network of the Noether school. Little wonder, then, that this success story has completely dominated nearly all the accounts of Noether's life. Most of Noether's publications from the first period, on the other hand, received little attention during her lifetime. This applies even to her famous paper "Invariant Variational Problems" [Noether 1918b], which today is perhaps her best-known single work. As documented in [Kosmann-Schwarzbach 2006/2011], this paper was rarely ever cited until many years after Noether's death. No doubt Weyl's periodization of her research interests offers a useful schema, so long as we are not misled by it into thinking that Emmy Noether's earlier work had little to do with her publications from the 1920s. As Uta Merzbach noted, a great deal of her work had clearly identifiable classical roots:

> Her deep knowledge of the literature and her ability to recognize and bring to the fore those concepts that would prove most fruitful prepared her . . . to undertake her grand synthesis. If one examines her work after 1910, one finds continual growth, but little change in methodological pattern. [Merzbach 1983, 169][9]

This should come as no surprise if we remember that Emmy Noether had a thorough knowledge of the mathematical literature of her time; she was well-versed in major works from the latter half of the nineteenth century.

As pointed out in [Koreuber 2015, 5], Weyl's tripartite framework is highly problematic if one hopes to gain a deeper understanding of Emmy Noether's intellectual growth. While our purpose here is certainly not to probe her works in any depth, in Chapter 2 we have attempted to give a balanced picture that restores the critically important first period in her career. Those years form part of the larger context taken up in Chapter 3, which deals with her life in Erlangen, where she grew up and was long known as the daughter of the eminent mathematician Max Noether. When Emmy Noether finally left Erlangen in 1915, she did so with the hope of joining the faculty in Göttingen, a plan supported by its two senior mathematicians, Felix Klein and David Hilbert. Their efforts, however, at first failed, and as recounted in Chapter 4, it took four long years before Noether was allowed to habilitate in Göttingen. During those years, both Hilbert and Klein had become deeply immersed in problems that stemmed from Albert Einstein's novel approach to gravitation, the general theory of relativity. Noether worked first with Hilbert and then with Klein, and in [Noether 1918b] she ultimately unraveled one of the major mathematical mysteries that they and Einstein had struggled to solve,

---

[9]A more recent study that argues for a similar view is [McLarty 2017].

namely, the role of energy conservation in physical theories based on variational principles.

Noether's best-known works in algebra stem from the second period, when she made major contributions to ideal theory. She was almost 40 when she published "Ideal theory in ring domains" [Noether 1921b], one of her most influential algebraic publications. Here she introduced the general concept of rings satisfying the ascending chain condition, familiar today as Noetherian rings. Soon afterward, her reputation as a leading algebraist began to spread beyond Göttingen, leading to her fame as "der Noether." In describing this work from her second period in Chapter 5, we sketch the general shift from classical to modern algebra illustrated by comparing Noether's work with Richard Dedekind's earlier theory of ideals in number fields.[10]

After this, the focus in Chapters 5, 6, and 7 turns to Noether's relationships with the four other mathematicians who appear in "Diving into Math with Emmy Noether": Bartel L. van der Waerden, Pavel Alexandrov, Helmut Hasse, and Olga Taussky. While none of these four took a doctoral degree under Noether, all were closely connected with her school in one way or another. Each, in fact, represents a strand of influence that ran through the Noether school, thereby contributing to its diverse and eclectic character. Van der Waerden's principal interests, when he arrived in Göttingen from Amsterdam, were closely related to Max Noether's work in algebraic geometry. After studying under Noether's daughter and then under Emil Artin in Hamburg, he published his classical two-volume textbook *Moderne Algebra* [van der Waerden 1930/31], which for decades afterward served as the standard introduction to the subject. The Russian topologist Pavel Alexandrov was a regular visitor in Göttingen during the summer months. As one of Emmy Noether's closest friends, he spent countless hours "talking mathematics" with her, eventually joined by another topologist, Heinz Hopf. These conversations proved of vital importance for the emergence of modern topology, a field that began to take on clear form in their textbook [Alexandroff/Hopf 1935]. Both van der Waerden and Alexandrov very consciously adopted Noether's conceptual approach in writing these two seminal works, which distilled and synthesized essential knowledge in two fundamentally new disciplines: abstract algebra and algebraic topology.

During the final phase of Noether's career, Helmut Hasse was her closest collaborator. As a student of Kurt Hensel in Marburg, Hasse developed a new local-global principle based on Hensel's p-adic numbers that proved highly fertile for research in algebraic number theory. As he began to explore a new research agenda for class field theory, Emmy Noether pointed out the relevance of ongoing work on hypercomplex number systems (i.e., non-commutative algebras) for generalizing the number-theoretic investigations of Hasse and Artin. Thanks to the carefully edited publication of her letters to Hasse, published in [Lemmermeyer/Roquette 2006], one can easily recognize how Noether's ideas had a catalytic effect on Hasse's work after 1927. Her constant, unrelenting prodding,

---

[10]For a detailed comparison, see [Corry 2017].

mixed with praise and encouragement, played a major part in their symbiotic relationship, underscoring the importance of purely human factors in mathematical research. Noether's parallel collaboration with Richard Brauer soon led to a threesome, who together succeeded in proving the Brauer-Hasse-Noether theorem.

Noether's relationship with Olga Taussky was unlike any other, not least because she, too, was a woman with a mind of her own. Their first lengthier interactions took place during the academic year 1931/32 when Taussky came to Göttingen as a young Viennese post-doctoral student, having been hired by Richard Courant to lend help in editing Hilbert's early works on number theory. She was highly qualified to do so, having studied under Phillip Furtwängler, a leading expert on class field theory. After returning to Vienna for two years, Taussky rejoined Emmy Noether at Bryn Mawr College in 1934, a difficult time in the lives of both women, as Taussky would recall late in her life. Olga Taussky never became an enthusiast for Noether's abstract style of mathematics, and yet her encounters with Emmy, particularly during the last year of her life, proved to be of great importance for the young woman's career.

Indeed, none of these four mathematicians – van der Waerden, Alexandrov, Hasse, and Taussky – who went on to write hundreds of papers and produce dozens of Ph.D.s in the course of their careers, can really be called a disciple of Emmy Noether, even though all of them were inspired by her ideas and personality in significant ways. Just this, in fact, explains why we want to focus attention on these four individuals. By highlighting their roles within the Noether school – a community quite unlike earlier, more conventional mathematical schools – we gain a clear sense of Emmy Noether's unique leadership style, her commanding presence, as well as the open-ended nature of this highly improvisational undertaking. This said, we do not wish to overlook the quite large group of those who were Noether's students in the narrower sense, in particular her special favorite Max Deuring. He and a few others receive some attention in our book as well, even though its principal aim is to spotlight the larger dimensions of the Noether school. As Hermann Weyl emphasized, she was at the very center of Göttingen mathematics, just as the Göttingen scientific community was a manifestation of Weimar Germany's vibrant cultural life. As one of Weimar culture's leading representatives, Albert Einstein later wrote about Emmy Noether's highly significant role in this ultimately tragic story.[11]

Chapter 8 describes the traumatic events of 1933 that dramatically ended mathematical life in Göttingen as Noether had known it. She and Richard Courant, the director of the Mathematics Institute, were both forced into exile in the United States. Helmut Hasse would ultimately be appointed to Courant's chair, but while still in Marburg he initiated a campaign to maintain Noether's modest position in Göttingen. Predictably, these efforts failed, but through the intercession of friends

---

[11] For the interpretation of Göttingen mathematics as a phenomenon within the larger context of Weimar culture, see [Rowe 1986]. Einstein's obituary of Noether is analyzed in Section 9.1 of *Emmy Noether – Mathematician Extraordinaire* [Rowe 2020b].

in the United States, Emmy Noether gained a temporary appointment at Bryn Mawr College, a distinguished institution of higher learning for women.

In Chapter 9, we briefly recount Bryn Mawr's importance for the history of mathematics before relating various events and circumstances connected with Noether's association with the college. Emmy Noether's last two years in the United States were filled with all kinds of worries, few of which she spoke about even with her closest friends. One of these was Anna Pell Wheeler, chair of the Mathematics Department at Bryn Mawr College, who in many ways helped her to adjust to life in the United States. During her 18 months there, she also began to spread the gospel of modern algebra in weekly lectures at Princeton's Institute for Advanced Study, where her seminar attracted a number of prominent as well as up-and-coming mathematicians. Her collaborator from Germany, Richard Brauer, attended regularly, as did Nathan Jacobson. The latter filled in for Emmy Noether at Bryn Mawr the following year and would later edit her *Collected Papers* [Noether 1983]. Her sudden death on 14 April 1935, following an operation, came as a huge shock to everyone, perhaps most of all to her brother Fritz, who also had been forced to leave Germany with his wife and two sons. Emmy had tried in vain to find work for him in the United States, after which he took a position at Tomsk Polytechnic University in Western Siberia.[12]

Among earlier scholars who studied the life and work of Emmy Noether, we would like to underscore the pioneering efforts of Auguste Dick, who gathered a good deal of interesting information in the process of writing [Dick 1970/1981]. These materials can be found today in her literary estate, located in the Archive of the Austrian Academy of Sciences in Vienna, which kindly granted us permission to reproduce several photographs that appear in this volume. Here we would like to mention an exchange of letters between Dick and B.L. van der Waerden, written shortly after she interviewed him in Zurich. On 28 February 1967, he asked whether he had understood her correctly regarding one of Dick's motivations in researching Noether's life, namely to refute the then standard view that women were less capable of doing creative work in mathematics than men. Van der Waerden found this general hypothesis unconvincing, but he also thought that he could reach an objective conclusion by means of a statistical test. In her reply, written on March 4, Dick, too, thought that a statistical analysis might shed new light on this question.

By this time, van der Waerden was steeped in studies of ancient science, guided in part by a widely shared view of Greek culture. In his letter to Dick, he offered the opinion that the ancient Greeks had a special genius for philosophy, mathematics, sculpture, and the composition of dramatic works. He also believed that modern-day Jews were similarly gifted, which seemed to suggest that Emmy Noether's Jewish background accounted for her intellectual brilliance. Being well versed in statistical methods, van der Waerden set up a null hypothesis based on

---

[12]The subsequent fate of Fritz Noether and his family is discussed in Section 9.2 of *Emmy Noether – Mathematician Extraordinaire* [Rowe 2020b].

the assumption that during the period from 1900 to 1950 mathematical talent was equally distributed between men and women.[13] He then sent the results to Auguste Dick in a letter from March 8; these, he thought, thoroughly refuted her contention. How she responded, if at all, we do not know. In the light of what we do know today, however, van der Waerden's mathematical exercise must strike any informed observer as hopelessly naive. Given the extremely limited professional opportunities open to women who chose to study higher mathematics during the early decades of the twentieth century, it would actually seem more pertinent to wonder why a surprisingly high number of doctoral degrees were conferred on females during this period.[14] Some of these women, particularly those who had an association with Göttingen University or Bryn Mawr College, appear elsewhere in this book, though we have only occasionally touched on the larger theme of roles for women in modern mathematics.

Any contextualized account of Emmy Noether's life cannot overlook the fact that she was a woman in a world totally dominated by men. She also happened to come from a Jewish family. In her reply to van der Waerden from March 4, Auguste Dick wrote that she was interested in Emmy Noether for a number of reasons, another being her Jewish background.[15] She understood from van der Waerden's remarks that he thought this a likely reason for her extraordinary talent, and Dick tended to agree. No doubt she was much surprised to learn that this was not what he meant at all. For he replied, saying that "Emmy Noether was unique and altogether different from all other Jews that I know. You probably know that she was called 'der Noether' in Göttingen. She was motherly, without being typically feminine, just as she was not typically Jewish."[16] Van der Waerden was here invoking familiar stereotypes, perhaps even ones that Noether shared as a product of those times.

A person is considered *discriminating* if they can discern significant differences; those who cannot, on the other hand, will often *discriminate* merely on the basis of personal prejudice. Within the small mathematical elite that Emmy Noether moved in, the latter mentality was anything but rare. Although their views on social or political affairs were by no means homogeneous, Noether's peers were often no more astute when it came to such matters than were "ordinary" people. Moreover, as suggested by van der Waerden's opinions relating to gender and race, such attitudes could easily affect a person's judgment when it came to evaluating other mathematicians' creative abilities.

---

[13]He based this test on the assumption that roughly 20% of those who studied at European or American universities were women.

[14]In [Green/La Duke 2009] the authors found that over 14% of the doctorates awarded in the United States during the period 1900–1940 were conferred on women.

[15]Dick's interest in the Noether family led her to undertake extensive genealogical research, as can be seen from several documents in her literary estate. She also researched the careers of mathematicians in Vienna and Prague who were persecuted during the Nazi era; see [Pinl/Dick 1974].

[16]This passage was cited in the original German in [Siegmund-Schultze 2011b, 216]. The letters from van der Waerden are in Nachlass-Dick 11-35, Archive of the Austrian Academy of Sciences in Vienna.

As we have noted in various places in this book, but especially in Chapter 8, there was a strong tendency in Germany to link mathematical style with ethnicity, a viewpoint by no means restricted to anti-Semites. In essence, this stemmed from the belief that traditional Talmudic studies had sharpened the minds of *male* Jews in specific ways that promoted logical thinking and critical acumen. As a result, so many believed, they excelled especially in those mathematical disciplines in which abstract theorizing played a major role, one of the most important being abstract algebra. Since Emmy Noether was the acknowledged leader of this direction in mathematical research, her work obviously fit this stereotype (disregarding her gender). Those who felt no sympathy for abstract algebra were easily inclined to regard this new trend as somehow "Hebraic," and hence foreign to what they imagined to be good, sound "German" mathematics.

These stereotypes were particularly widespread in Germany during the postwar years of the Weimar Republic, a period during which career opportunities for Jewish mathematicians improved markedly. And although such views engendered little serious discussion before 1933, much less open debate, afterward they passed smoothly into more familiar forms of Nazi propaganda. Faced with seeing their beloved teacher banned from the German universities, Noether's faithful students countered by underscoring how her work was anchored entirely in the tradition of Richard Dedekind, one of the great German mathematicians of the nineteenth century. To follow the logic of their argument, they as good Germans wanted the government to recognize that Noether was working on behalf of "Aryan mathematics." Their efforts were in the end in vain, but the thrust of all this stemmed evidently from the conviction that "most" German Jews had a distinctly different way of thinking about mathematics, one that favored abstract theorizing while neglecting fields with close ties to the physical sciences. No one, it seems, whether then or earlier, seriously questioned that opinion by pointing out that for every Emmy Noether representing the first tendency, there was a Fritz Noether practicing the second. This larger point was brought out forcefully in the exhibition "Transcending Tradition: Jewish Mathematicians in German-Speaking Academic Culture," which presented a wide array of works by German-Jewish scholars. These completely refute the claim that there was a "typical form of 'Jewish mathematics', remote from geometrical intuition or from applications" [Bergmann/Epple/Ungar 2012, 134].[17]

A book such as this one could obviously not have been written without the efforts of many others, including those whose names appear in the works cited throughout. Rather than attempting to discharge our debt by individually listing all those who have contributed directly or indirectly to making this book possible,

---

[17]David Hilbert was one of the few who forthrightly claimed that "mathematics knows no races" (see [Siegmund-Schultze 2016]). His colleague, Felix Klein, thought that geometrical intuition (*Anschauung*) was deeply rooted in the Teutonic race. Yet as Klein and Max Noether well knew, after 1890 this impulse lost ground in Germany just as it was being taken up by leading Italian geometers: Corrado Segre, Guido Castelnuovo, and Federigo Enriques, all of whom were of Jewish descent.

we take this opportunity to thank everyone collectively, while singling out only a few who deserve special thanks. We must begin, first of all, by thanking the representatives of the Berlin Mathematics Research Center MATH+ for their vital support in making the interdisciplinary conference in honor of Emmy Noether possible. This was a truly unique multi-disciplinary event, one that led to many fruitful discussions and intellectual exchanges from a variety of perspectives. For those involved in its conception and realization, it proved to be a rewarding experience that also offered many new reflections with regard to Emmy Noether's past and present relevance for mathematics. Second, our thanks go to the aforementioned Sandra Schüddekopf and Anita Zieher of *portraittheater Vienna*, along with the four actors who appear in supporting roles: Alexander E. Fennon (van der Waerden), Werner Landsgesell (Alexandrov), Helmut Schuster (Hasse), and Karola Niederhuber (Taussky). Their collective efforts provided the true inspiration for this book. We are also grateful for the cooperation we received from archivists working at the Austrian Academy of Sciences, Bryn Mawr College, Caltech, Göttingen State and University Library, and Oberwolfach Research Institute for Mathematics. All provided essential help to us in procuring documents and digital images for this book. In particular, we also wish to thank Qinna Shen, Professor of German Studies at Bryn Mawr College, for her efforts in supporting this project. Thanks also to Ayse Gökmenoglu for the care she took in producing the photos included in it, and to Catriona Byrne and Rémi Lodh at Springer Nature for their helpful advice in supporting this venture.

Among those who read and commented on parts of the text at some stage, we wish to thank Leo Corry, Joe Dauben, Hilde Rowe, Manfred Lehn, Jemma Lorenat, Monica Noether, Volker Remmert, Peter Roquette, Erhard Scholz, Reinhard Siegmund-Schultze, Margaret Noether Stevens, Evelyn Noether Stokvis, and Cordula Tollmien. Finally, we owe special thanks to Walter Purkert, former coordinator of the Hausdorff editorial project and coauthor with Egbert Brieskorn of [Brieskorn/Purkert 2018], the monumental Hausdorff biography. Walter read the entire manuscript and offered us a number of very helpful ideas for our interpretation of Alexandrov's role in the larger story told here. So while this book began initially as a companion to "Diving into Math," our hope is that it will not only serve to promote the play but also to spark renewed interest in Emmy Noether's remarkable life and her fascinating mathematical world.

David E. Rowe
Mechthild Koreuber

Chapter 1

Diving into Math with Emmy Noether

Some 100 years ago a notice appeared in the journal of the German Mathematical Society that read: "Dr. Emmy Noether has habilitated as a lecturer in mathematics at Göttingen University." This quiet announcement was actually the resounding final chord in a long struggle that went on for four years and only ended on June 4, 1919, when Noether joined the Göttingen faculty. Habilitation, which long stood as a formal bulwark that prevented women from teaching at German universities, had finally fallen. For Emmy Noether, this new status represented a marvelous opportunity to pursue her mathematical interests in research and teaching without any restrictions or limitations. A distinctive approach to mathematics – one that placed abstract algebraic concepts and their mutual relationships at the center of research – would soon become Noether's credo. Her aim was to show how this approach could lead to an enrichment of mathematics, a goal she shared with the many talented young mathematicians who by the early 1920s had begun to gather around her.

Anniversaries mark an occasion to reflect on a person's life and work, whereas a theatrical production can offer the illusion of bringing someone back to life (Fig. 1.2). In our play we made use of correspondence, obituaries, and the memories of Emmy Noether's contemporaries in an attempt to approach her historical figure through documentary sources, while allowing her personality to unfold on stage. Noether's life was complicated, with many ups and downs. Despite her outstanding mathematical achievements, she experienced professional discrimination and marginalization, not to mention the racist threats that forced her to emigrate. Yet her life was also simple, because throughout it mathematics formed the center of her interests. Despite all the adversity and rejection, Emmy Noether's passion for mathematics never wavered and this guided her through an extraordinary life.

D. E. Rowe, M. Koreuber, *Proving It Her Way*, https://doi.org/10.1007/978-3-030-62811-6_1

1.1 Emmy Noether's Timeline

1882 Amalie Emmy Noether is born on March 23 as the first child of Jewish parents in Erlangen. Her father, Max Noether, is professor of mathematics at the university; her mother, Ida Noether née Kaufmann, comes from a wealthy family in Cologne. Emmy Noether will eventually have three younger brothers.

1889–1897 she attends the secondary school for girls in Erlangen.

1900 after private tutoring, she passes the Bavarian examination for French and English language teachers. Her goal, though, is not to work as a teacher, but to pursue her academic interests.

1900–1903 she attends courses in mathematics, romance languages, and history at Erlangen University, while preparing to take the qualifying exam required for entry, which she passes in 1903.

1903/04 as a guest auditor in Göttingen she attends mathematics courses taught by David Hilbert, Felix Klein, and others. By the end of the semester she becomes ill and returns home. The following year, she enters Erlangen University as a regularly enrolled student of mathematics.

1907 she graduates *summa cum laude* with a dissertation written under Paul Gordan on the invariants of a ternary biquadratic form. The work reflects her mentor's algorithmic approach to invariant theory. Later she will describe it as "calculations," "formula thickets," or even just as "crap."

1907–1915 after graduation Noether continues to work on mathematics in Erlangen. Although unpaid, she works voluntarily to support courses taught by her father and Gordan. After Gordan's retirement, she does the same under his successors, Erhard Schmidt (1910) and Ernst Fischer (from 1911). During walks and discussions with Fischer, she gains her first insights into abstract algebra.

1909 she becomes a member of the German Mathematical Society (Deutsche Mathematiker-Vereinigung) and delivers a lecture at its annual meeting in Salzburg, the first woman to do so. For years afterward she would regularly speak at these meetings, interacting with other mathematicians, and taking an active part in the discussions that followed their talks.

1913 she accompanies her father to Göttingen, where they meet with Klein to discuss plans for the obituary Max Noether will write for his colleague Paul Gordan. Klein is deeply impressed by the breadth of her mathematical knowledge.

1915 she receives an invitation from Klein and Hilbert to pursue post-doctoral studies in Göttingen. Hilbert engages her to help promote his research program in field physics based on Einstein's general theory of relativity. Noether's expertise in invariant theory makes her an ideal assistant for Hilbert's research work. With the support of the Göttingen mathematicians, she applies for certification to teach, submitting a post-doctoral (habilitation) thesis connected with Hilbert's 14th Paris problem. Her application leads to an intense controversy in the faculty, which votes to refer the matter to the Ministry of Culture. Since this case involved a fundamental question – whether a qualified woman had the right to teach – the Ministry refuses to set this precedent. Instead, a compromise is reached enabling Noether to teach courses announced under Hilbert's name.

1918 Klein submits her paper on differential invariants to the Göttingen Scientific Society. Einstein writes to Hilbert: "It impresses me that one can view these things from such a general standpoint. It wouldn't have hurt the Göttingen troops in the field if they had been sent to Frl. Noether."

1919 in the wake of the profound social and political changes in Germany after the First World War, women gain the right to teach at universities. Noether submits a new habilitation thesis, her famous paper "Invariant Variational Problems," already published in 1918. Its two main theorems show the connection between symmetries and conserved quantities in variational systems under very general conditions. One of its main applications clarifies the status of energy conservation in the general theory of relativity. Not until the 1950s, however, will theoretical physicists come to appreciate the importance of these theorems, which today bear her name.

4 June 1919 Emmy Noether presents her inaugural lecture on module theory as the final requirement before receiving the *venia legendi* and joining the Göttingen faculty. With her new status, she now begins her teaching career as a Privatdozent, a position without remuneration.

1921 she publishes one of her seminal papers, "Idealtheorie in Ringbereichen," which adopts the abstract approach for which she will become famous. Written with unusual clarity, this classical contribution to the ideal theory of "Noetherian rings" can still be read today by students wishing to learn key concepts in abstract algebra. Over the next several years, ideal theory would remain her principal field of research.

1922 Noether receives the honorary title of associate professor, a distinction signifying recognition, but without any financial benefits.

1923 she gains for the first time a teaching appointment that provides a small remuneration to support her scientific work. Although living very modestly after her father's death in 1921, she finds herself in dire economic straits due to the rampant hyperinflation.

1926 Grete Hermann takes her doctoral degree as the first official graduate of the "Noether school." In addition to Hermann, Noether's pupils include Max Deuring,

Heinrich Grell, Jakob Levitzki, and Kenjiro Shoda. A number of prominent foreign mathematicians, in particular Bartel L. van der Waerden from Amsterdam and Pavel Alexandrov from Moscow, come to Göttingen around this time. A close, family-like atmosphere pervades the fledgling "Noether school," composed of young mathematicians eager to learn about abstract algebra and its applications to other mathematical disciplines.

1928/29 Noether accepts an invitation from Alexandrov to spend a semester in Moscow as a visiting professor. She offers a lecture course on modern algebra attended by 20-year-old Lev Pontryagin, who became one of the century's leading topologists. On returning to Göttingen, she speaks positively about social developments in the Soviet Union. Her sympathy for the Bolshevik experiment soon earns her the reputation of being a Marxist.

1929 although initially viewed with skepticism by some of her Göttingen colleagues, Noether's research and that of her students reveals the fruitfulness of her methodological approach. A highly communicative oral style becomes her trademark and her lecture courses take on programmatic significance. Other young mathematicians – including Helmut Hasse, Gottfried Köthe, and Chiungtze Tsen – take up her ideas and develop them further.

1930 she teaches for one semester in Frankfurt, filling in for Carl Ludwig Siegel during his leave of absence. One of her students there is Paul Dubreil, later a member of the Bourbaki Circle. He and his wife, Marie-Louise Dubreil-Jacotin, attend Noether's lectures in Göttingen. They both go on to successful itinerant careers in France, where Dubreil-Jacotin becomes the first female professor of mathematics.

1932 Noether receives the Ackermann–Teubner Memorial Prize together with Emil Artin. At the International Congress of Mathematicians in Zurich she gives a keynote lecture on hypercomplex systems in relation to commutative algebra and number theory. Despite such distinctions and international recognition, she is never seriously considered for a professorship.

25 April 1933 she and several of her colleagues in mathematics are persecuted by the Nazi regime, which enacts a Law for Restoration of the Civil Service. Although she was never a civil servant, she is placed on temporary leave while her case is under review. Colleagues and students try to support her with testimonials and petitions, but on September 2, 1933 her teaching license is revoked on the grounds of her Jewish ancestry.

1933 she receives an invitation from Somerville College in Oxford, but also an offer for a visiting professorship at Bryn Mawr College in Pennsylvania. When the financial arrangements in Oxford remain cloudy, she accepts the offer from Bryn Mawr.

1934 in addition to her courses at Bryn Mawr, Noether gives weekly lectures on topics in algebra at the nearby Institute for Advanced Study in Princeton. There she encounters several familiar figures, fellow émigrés like Hermann Weyl and his assistant, Richard Brauer. In the summer Noether travels to Germany and visits her brother Fritz, who will soon emigrate to the Soviet Union. She lectures in Hamburg and disposes of her household belongings in Göttingen before returning to Bryn Mawr. There she is joined by another émigré, Olga Taussky, who becomes part of a new circle of "Noether girls."

14 April 1935 Emmy Noether dies from complications following an operation for the removal of a fibroid. Her ashes are placed in an urn and buried under the cloisters of the M. Carey Thomas Library on the campus of Bryn Mawr College.

Figure 1.1: Anita Zieher as Emmy Noether, Courtesy of Portraittheater Vienna

(a) Emmy Noether, ca. 2020

(b) Emmy Noether, ca. 1930

Figure 1.2: Emmy Noether and her Doppelgängerin, Portraittheater Vienna and Emmy Noether Papers, Bryn Mawr College Special Collections

1.2   Four Mathematicians from Noether's Circle

Bartel L. van der Waerden (1903–1996) was a Dutch mathematician whose research interests ranged from abstract algebra and algebraic geometry to quantum mechanics. In 1919 he began studying mathematics in Amsterdam; one of his teachers, L.E.J. Brouwer, recommended him to Emmy Noether. During his first stay in Göttingen in 1924/25, he learned modern algebra from her. Inspired by these studies, he applied algebraic methods in his doctoral dissertation on the foundations of enumerative geometry. He then returned to Göttingen on a Rockefeller fellowship, which also took him to Hamburg. There he and Emil Artin considered writing a textbook on modern algebra, which van der Waerden later published on his own. After spending four years in Groningen, van der Waerden was appointed professor of mathematics in Leipzig in 1931. After the war, he held a professorship in Amsterdam and then at the University of Zurich, where he taught from 1951 until his retirement in 1973.

With the publication of his two-volume textbook *Moderne Algebra* in 1930–31, van der Waerden made a decisive contribution to the canonization of abstract algebra. Based in part on lectures by Noether and Artin, these volumes played a major role in promoting a deeper understanding of and appreciation for the new structural approach to mathematics.

Pavel Alexandrov (1896–1982) was a Soviet mathematician who did pioneering work in topology. During his studies at Lomonosov University in Moscow he met Pavel Urysohn, and they soon became intimate friends. In 1923 they traveled to Göttingen, where they interacted with several mathematicians, in particular

Emmy Noether. Their passionate enthusiasm for topology matched well with her equally strong interest in new abstract directions in mathematics.

Following Urysohn's tragic death in 1924, Alexandrov visited Göttingen often to discuss mathematics with Noether. Her ideas helped stimulate the development of algebraic topology, as reflected in *Topologie* (1935), the classic text by Alexandrov and Heinz Hopf. Alexandrov spent the academic year 1927/28 in Princeton, and in 1928 he was named corresponding member of the Göttingen Academy of Sciences. At his invitation, Noether spent the winter semester of 1928/29 in Moscow. Her inspiring lectures on abstract algebra were followed there with great interest. Soon after she returned to Göttingen, Alexandrov was appointed to a professorship at Lomonosov University, but once the Nazi government came to power he no longer traveled to Germany.

Helmut Hasse (1898–1979) was a German mathematician who specialized in algebraic number theory and abstract algebra. After WWI he began his studies in Göttingen, but left in 1920 to study under Kurt Hensel in Marburg. He taught in Kiel before his appointment to a professorship in Halle in 1925. In 1930, Hasse succeeded Hensel in Marburg and four years later he was appointed to the professorship in Göttingen formerly held by Hermann Weyl before he resigned. Despite the fact that she was living in exile, Emmy Noether congratulated him on this appointment and expressed her hopes that Göttingen would remain an important mathematical center. After the war, the British authorities in Göttingen decided that Hasse could not be reappointed in view of his past support of the Nazi regime. He thereafter accepted a research position at the Academy of Sciences in East Berlin, and in 1949 took on a professorship at the Humboldt University. One year later, he left this position to accept a chair at the University of Hamburg, where he remained until his retirement in 1966.

Hasse made fundamental contributions to algebraic number theory, particularly class field theory and the theory of algebras. He and Noether engaged in an intensive correspondence that clearly conveys Noether's ability to inspire creative mathematical exchanges. In 1932, Hasse, Noether, and Richard Brauer published one of the central theorems in the theory of algebras.

Olga Taussky (1906–1995) was an Austrian-American mathematician whose early work centered on algebraic number theory. After taking her doctorate in Vienna under Philipp Furtwängler, Richard Courant invited her to Göttingen in 1931 so that she could assist with the editing of the first volume of David Hilbert's Collected Works. During this stay she attended Emmy Noether's lectures and struck up a friendship with her. In 1934 Taussky was awarded a scholarship from Bryn Mawr College, which gave her the chance to work with Noether again.

After Emmy's Noether's death in 1935, Taussky became a Fellow at Girton College in Cambridge University. She married the Irish mathematician, Jack Todd, and both undertook war-related research before coming to the United States in

1945. In 1957, they moved to Caltech, where he held a professorship, whereas she was employed only as a research assistant with the same obligations. Although the author of over 300 research papers, it was not until she was approaching retirement in 1971 that she was finally promoted to a professorship. In 1981 she delivered the annual Noether Lecture, an honor the Association for Women in Mathematics bestows on distinguished female mathematicians.

1.3   Name Dropping: A Who's Who

A. Adrian Albert (1905–1972), leading American algebraist in the tradition of the Chicago school, his work on finite-dimensional division algebras linked with that of the German algebraists in proving the Albert–Brauer–Hasse–Noether theorem; from 1931 professor University of Chicago, 1934 guest researcher Institute for Advanced Study in Princeton, 1956/57 President of the American Mathematical Society.

Emil Artin (1898–1962), Austrian algebraist and one of the century's leading figures in algebraic number theory; 1925–1937 professor University of Hamburg, 1927 solved Hilbert's 17th problem in "Decomposition of positive definite functions into squares," 1932 gives famous lectures on class field theory attended by many in Göttingen, including Noether and Olga Taussky, recipient of the Ackermann-Teubner Memorial Prize together with Emmy Noether; 1937 emigration to the USA (Notre Dame), 1938 Indiana University, 1946 Princeton University, 1958 return to University of Hamburg.

Richard Brauer (1901–1977), German-American algebraist and collaborator of Emmy Noether, studied under Isaai Schur in Berlin, private lecturer in Königsberg until removal by Nazi government; 1933 assistant professor at University of Kentucky, 1934 assistant to Hermann Weyl at Princeton's Institute for Advanced Study, 1935 professor University of Toronto on recommendation of Emmy Noether, 1948–1952 University of Michigan, 1952–1971 Harvard University.

L.E.J. Brouwer (1881–1966), Dutch topologist and a leading mathematician of his generation, pioneer of geometric topology who proved invariance of dimension, founder of intuitionism opposing Hilbert's formalism in the philosophy of mathematics; from 1909 professor at the University of Amsterdam, Bartel L. van der Waerden among his students from 1919 to 1923, close ally of Emmy Noether, 1914–1928 member of editorial board of *Mathematische Annalen*, abrupt dismissal in December 1928 ended his ties with Göttingen.

Richard Courant (1888–1972), German-American analyst, student of David Hilbert, 1921 professorship in Göttingen, 1922–1933 director of Mathematical Institute, 1924 publication of Courant & Hilbert *Methoden der mathematischen Physik*; forced to step down as director, takes chair at New York University, builds up graduate program in applied mathematics relying heavily on European refugees, after WWII returns to Göttingen nearly every summer, 1964 opening of the Courant Institute of Mathematical Sciences, today one of the leading research centers in the world.

Richard Dedekind (1831–1916), last living representative of the old Göttingen school, 1852 doctorate under Carl Friedrich Gauss, 1854 habilitation soon after his friend Bernhard Riemann, 1855–1858 works closely with Gauss's successor Gustav Lejeune Dirichlet; 1858 professor Zurich Polytechnic and from 1862 Brunswick Institute of Technology, influential publications include 11th supplement to Dirichlet's lectures on number theory and Dedekind's works on ideal theory, frequently used by Noether in courses marking the beginning of abstract algebra.

Max Deuring (1907–1984), Emmy Noether's favorite student, his 1930 dissertation on the arithmetic theory of algebraic functions received her highest praise; 1931–37 van der Waerden's Assistent in Leipzig, 1937–43 Privatdozent in Jena, 1935 published report entitled *Algebren* with Noether's support; after WWII held professorships in Marburg and Hamburg, 1950–1976 professor in Göttingen.

Albert Einstein (1879–1955), German-Swiss-American physicist and founder of theory of relativity, 1915 six lectures on the theory of relativity in Göttingen at invitation of Hilbert; 1916-1918 learns of Noether's work, 1915-1933 corresponding member of the Göttingen Academy of Sciences, 1918 supports Noether's candidacy for habilitation; 1921 Nobel Prize in Physics, 1933 emigration and research professorship at Princeton's Institute for Advanced Study, 1935 obituary for Noether in the *New York Times*.

Ernst Fischer (1875–1954), Austrian mathematician who studied in Vienna with the invariant-theorist Franz Mertens, remembered today for work of 1907 on the Riesz-Fischer Theorem; from 1902 worked at Brünn Institute of Technology, 1911 professor in Erlangen as successor to Gordan and from 1920 in Cologne, 1931 president of the German Mathematical Society, 1938 forced to retire due to Jewish ancestry, Fischer gave Emmy Noether, in her own words, "the decisive impulse to pursue research in abstract algebra."

Paul Gordan (1837–1912), famous as "king of invariant theory" for proving finite basis theorem for invariants of binary forms, remembered as an algorist who initially opposed Hilbert's groundbreaking work on invariant theory; 1863–1868 close collaboration with Alfred Clebsch at University of Giessen, from 1874 University of Erlangen, collaborator with Felix Klein, from 1875 colleague of Max Noether, 1907 Ph.D. supervisor of Emmy Noether, who later lost interest in the symbolic calculus her former mentor exploited so skillfully.

David Hilbert (1862–1943), leading mathematician of his generation whose early work revolutionized invariant theory, founder of formalist philosophy and opponent of Brouwer's intuitionist views; 1892 professor in Königsberg and from 1895 in Göttingen, 1900 programmatic address at the Second International Congress of Mathematicians with 23 "Hilbert problems," problem 14 on invariant theory inspires Emmy Noether's first Habilitation thesis, 1915–1917 pursues unified field theory combining Einstein's gravitational theory with Gustav Mie's theory of matter, assisted by Noether, who in 1918 derives Hilbert's central heuristic theorem.

Heinz Hopf (1894–1971), leading geometer and topologist, studied in Breslau and Berlin, proved the Poincaré-Hopf theorem; 1925/26 as post-doc in Göttingen meets Noether and Alexandrov, 1927/28 at Princeton University on a Rockefeller fellowship begins collaboration with Alexandrov on their monograph *Topologie*; 1931 succeeds Weyl at the ETH in Zurich, 1935 publication of Alexandov & Hopf *Topologie*, 1955–1958 president of the International Mathematical Union.

Felix Klein (1849–1925), mathematician with universal interests famous for his "Erlangen program," architect of modern Göttingen mathematical community; 1872 professor Erlangen, 1875 Munich Institute of Technology, 1880 Leipzig, from 1886 in Göttingen, editor-in-chief of *Mathematische Annalen*, 1897, 1903 and 1908 President of the German Mathematical Society, 1914 first winner of the Ackermann-Teubner Memorial Prize; 1917–1918 works closely with Emmy Noether on mathematical problems related to Einstein's general theory of relativity.

Sonya Kovalevskaya (1850–1891), Russian mathematician who from 1870–74 became Karl Weierstrass' favorite pupil in Berlin, 1874 doctorate *in absentia* from Göttingen; 1884 Stockholm's *Högskola* makes her the world's first female professor of mathematics, member of editorial board of *Acta Mathematica*, 1888 winner of the Prix Bordin of the French Academy, from 1889 corresponding member of the Russian Academy of Sciences; after her tragic death in 1891 followed posthumous fame based on two literary works, her own *A Russian Childhood* and a sequel biography written by her friend Anne Charlotte Leffler.

Edmund Landau (1877–1938), German expert on analytic number theory, studied in Berlin, 1901–1909 Privatdozent University of Berlin; 1909 appointed as successor of Hermann Minkowski in Göttingen, 1915 took a mixed view of Noether's candidacy for Habilitation: "How easy this decision would be for us if this were a man with exactly these accomplishments," but soon came to appreciate her merits; a proud Zionist supported founding of Hebrew University, where he taught in 1927, November 1933 student boycott of his lecture course led to international outcry.

Ruth Stauffer McKee (1910–1993), American mathematician who studied at Bryn Mawr as Emmy Noether's last Ph.D. student, after the latter's death examined by Richard Brauer; taught at Bryn Mawr School in Baltimore while undertaking postdoctoral work with Oscar Zariski, worked 30 years as a statistical analyst for the Joint State Government Commission in Harrisburg, Pennsylvania.

Fritz Noether (1884–1941), brother of Emmy Noether, specialized in applied mathematics and mechanics beginning with studies in mathematics and physics under Arnold Sommerfeld in Munich; 1910 completes volume 4 of Klein & Sommerfeld *Theorie des Kreisels* (Theory of the Top), 1911 habilitates at Karlsruhe Institute of Technology, 1922 professor at Breslau Institute of Technology, 1934 emigration to the Soviet Union and professorhip at University of Tomsk; arrested in 1937 for alleged espionage and executed in 1941, rehabilitated in 1988 by the Supreme Court of the Soviet Union.

Max Noether (1844–1921), regarded as one of the founders of modern algebraic geometry along with his mentor Alfred Clebsch, father of Emmy Noether; from 1875 professor in Erlangen, 1899 President of the German Mathematical Society, 1893 coauthor with Alexander Brill of classic survey of older and newer research on algebraic functions, author of numerous scientific obituaries for distinguished mathematicians associated with *Mathematische Annalen* including his colleague Paul Gordan; Noether's impressive knowledge and extraordinary care with respect to citations and historical accuracy are reflected in Emmy Noether's fastidious research standards.

Grace Shover Quinn (1906–1998), American mathematician, 1934/35 studied at Bryn Mawr on Emmy Noether Scholarship, 1937–1942 instructor Carleton College, 1956–1970 professor at American University, Washington D.C.

Ernst Steinitz (1871–1928), a distinguished geometer and algebraist who studied in Breslau and Berlin, wrote fundamental works on configurations and polyhedra as well as two survey articles on these topics in the German Encyclopedia; taught in Charlottenburg and Breslau, from 1920 professor at University of Kiel, his paper on the abstract theory of fields (1910) deeply influenced Emmy Noether's research.

Pavel Urysohn (1898–1924), gifted Russian who did pioneering work in topology, 1923/24 with his friend Pavel Alexandrov visits Felix Hausdorff in Bonn, L.E.J. Brouwer in the Netherlands, and the mathematicians in Göttingen, in particular Emmy Noether, loses his life when caught in a storm while swimming in the Atlantic.

Oswald Veblen (1880–1960), American geometer, graduate studies at University of Chicago with Ph.D. in 1903 under E.H. Moore, 1905–1932 professor Princeton University, after 1933 member of mathematics faculty Institute for Advanced Study; 1932 met Noether during stay in Göttingen, later he supported plan to create permanent position for her at IAS.

Marie Weiss (1903-1952), American algebraist who specialized in group theory, 1925 B.A. Stanford, 1926 M.A. Radcliffe College, 1928 Ph.D. Stanford with a dissertation published in *Transactions of the American Mathematical Society*; from 1930 assistant professor at Newcomb College, Tulane University, 1934/1935 in Bryn Mawr on an Emmy Noether Fellowship, from 1938 professor at Newcomb College, 1949 publication of her textbook *Higher Algebra for the Undergraduate*.

Hermann Weyl (1885–1955), leading German mathematician of his generation with broad research interests in analysis, mathematical physics, foundations and philosophy of mathematics; 1913 professor at the ETH Zurich, 1930 Hilbert's successor and Noether's colleague in Göttingen, 1932 President of the German Mathematical Society, 1933 emigration to the USA joining the faculty of the Institute for Advanced Study in Princeton, where Noether regularly lectured, April 1935 Memorial Address for Emmy Noether at Bryn Mawr College.

Anna Pell Wheeler (1883–1966), studied mathematics at University of South Dakota, taking master's degrees there and at Radcliffe College, before beginning doctoral studies in Göttingen, 1909 Ph.D. University of Chicago under E.H. Moore; 1911–1918 taught at Mount Holyoke College, 1918 associate professor Bryn Mawr College, 1925–1948 full professor and head of department; she played a key role in bringing Emmy Noether to Bryn Mawr, working and socializing in harmony until the latter's death.

Ernst Witt (1911–1991), German algebraist, studied in Göttingen under Emmy Noether until her dismissal, 1934 Witt takes his Ph.D. with Gustav Herglotz substituting for Noether; 1936 habilitation in Göttingen with Helmut Hasse, from 1938 professor at University of Hamburg, interrupted by the war years and the denazification process conducted by British occupation authorities.

Chapter 2

Emmy Noether: a Portrait

"I always went my own way in teaching and research," Emmy Noether once wrote toward the end of her life.[1] This brief remark at once reflects not only her approach to mathematics but also the basic tone of her biography. Noether was one of the most important mathematicians of the twentieth century. Through her research and teaching she made a decisive contribution to the development of modern algebra as well as to the emergence and spread of the structural approach to mathematics. At the same time, she was a visionary figure with a passionate belief in the fertility of abstract concepts, not only for clarifying concrete problems but for building general theories that go far beyond the classical theories of the nineteenth century.

In his obituary for her, Bartel L. van der Waerden described Noether's unique and powerful personality in these words:

> We have found these to be her outstanding characteristics: the ability to tirelessly and consistently pursue conceptual penetration of her subject matter in order to achieve utmost methodological clarity; a tenacious insistence on methods and concepts she found valuable, no matter how abstract and unproductive they might appear to her contemporaries; and the aspiration to arrange all specific connections within particular, general conceptual schemata. In a number of respects, her thinking does indeed differ from that of most other mathematicians. We all gladly lean on figures and formulas. For her, these aids were worthless or even disruptive. She only worked with concepts, not intuition or calculation. [van der Waerden 1935, 476]

Emmy Noether's extraordinary mathematical gifts were matched by her boisterous, extroverted, and occasionally aggressive personality. No one ever described her as an "elegant lady," quite the opposite. Even at an early age, she took no heed

---

[1]In a letter from January 30, 1935, to Edward R. Murrow, then Assistant Secretary for the Emergency Committee in Aid of German Scholars [Siegmund-Schultze 2009, 380].

D. E. Rowe, M. Koreuber, *Proving It Her Way*, https://doi.org/10.1007/978-3-030-62811-6_2

of dressing according to the fashions of her day. In 1913, when she attended a conference in Vienna, the thirty-one-year-old *Mathematikerin* received an invitation from Professor Franz Mertens to visit him at his home. Decades later, Mertens' grandson had a vivid recollection of this exotic guest, who looked nothing like any other woman he had ever seen. To him, Noether appeared "like a Catholic chaplain from a rural parish – dressed in a black, almost ankle-length, and rather nondescript coat, a man's hat over her short hair (still a rarity at that time) and with a shoulder bag carried crosswise like those of the railroad conductors during the imperial period – she was a rather odd figure" [Dick 1970/1981, 1981: 22]. Numerous (male) mathematicians commented on Emmy Noether's physical appearance, often disparagingly or, in the case of Hermann Weyl, by way of invidious comparison with the other female mathematician whose name has often been linked with Noether's, namely Sofia Kovalevskaya. Weyl's views, like those of van der Waerden cited at the end of our preface, reflect certain standard prejudices from this era regarding women in mathematics.

Hermann Weyl regarded Emmy Noether as a mathematician whose genius transcended the bounds of her sex (see Section 9.6). The mathematicians in Göttingen spoke of her "in awed mockery" in the masculine form as "der Noether", an epithet that was not generally meant kindly. "She was heavy of build and loud of voice, and [it was difficult] to get the floor in competition with her. She preached mightily and not as the scribes. She was a rough and simple soul, but her heart was in the right place." Other remarks suggest that Weyl saw Emmy Noether as arrested in her development as a woman; she was, for him, "an unwieldly child" who lacked "essential aspects of human life," among these, he supposed, was the erotic dimension. Kovalevskaya, on the other hand, was "certainly the more complete personality but was also of a much less happy nature" [Weyl 1968, 3: 443]. This "unhappiness" trope was by then firmly established in accounts of the Russian woman's conflicted life.[2] This made Sofia Kovalevskaya far more interesting for Weyl, who claimed that "mathematics made her unhappy, whereas Emmy found the greatest pleasure in her work" (*ibid.*). The larger implication would seem clear: if a woman wants to pursue a life in mathematics, she will have to give up certain things; such a life demands sacrifices, and these diminish her chances of becoming a "complete woman." In Weyl's eyes, Emmy was willing to make those sacrifices, whereas Sonya could not.

Van der Waerden would probably have balked at drawing such a sweeping conclusion, if only due to the evident problem of sample size (statisticians, after all, require a bit more than two test cases). What seems so striking today is how freely Weyl engaged in this amateur psychologizing without paying the slightest attention to the social and educational hindrances that stood in the way for any woman with the talent and ambition to pursue a career as a research mathematician. Beyond these glaring inequities and the fact that Kovalevskaya and Noether

---

[2] Alice Munro turned this motif upside down in her short story "Too Much Happiness"; for an analysis of how this theme came into prominence shortly after Sofia Kovalevskaya's death in 1891, see [Kaufholz-Soldat 2019].

were among the first who managed to break through such formidable barriers, there would seem to be very little basis for comparing these two women *as mathematicians*. The former was an analyst, the latter an algebraist; Sonya applied Weierstrassian methods to solve special problems, whereas Emmy generalized theories that Dedekind, Hilbert, and others had developed before her. There is no evidence to suggest or reason to believe that Emmy Noether ever looked at any of Sofia Kovalevskaya's papers,[3] which reflected ideas from a wholly different mathematical world. In "Diving into Math," Emmy tells us just how different: if she composed mathematics that somehow reminds us of Johann Sebastian Bach, then Sonya's math might just as well be compared with the music of Louis Armstrong.

2.1 Biographical Aspects

Emmy Noether's birthplace, the city of Erlangen in northern Bavaria, was by the end of the 19th century shaped by its well-to-do bourgeoisie, a flourishing economy, and especially by its university. Her father, Max Noether, was one of the university's two professors of mathematics, along with his older colleague Paul Gordan. Emmy Noether, Max and Ida Noether's eldest child, grew up with her three brothers in a liberal atmosphere typical of recently integrated Jewish families. In this milieu, one took for granted that daughters received a proper education, and so it was with Emmy. She attended a secondary school for girls, and then went on two years later to pass the state examination for teachers in French and English in 1900. The Noether family – Max and Ida and their four children – were members of the local Jewish community, which numbered around 200 persons during Emmy's childhood, less than 2% of the city's population. More important still, all were recent arrivals, as before 1861 Jews were not permitted to live within the city limits. Max Noether and Paul Gordan were both Jewish by birth, which made Erlangen unique within the German mathematical community, especially given the small number of university professorships available throughout Germany. Given the pervasive character of antisemitism in German academia, we will take a closer look at how this form of discrimination affected the careers of Gordan and Max Noether in Section 3.2.

A far more blatant form of institutionalized discrimination involved the two sexes, as only graduates of more elite types of schools could enter a university, and girls were not allowed to attend these. Thus, young women had virtually no chance of studying at a university, let alone dreaming of a teaching career at one of these institutions. That Emmy Noether dreamed of such a life at an early age probably cannot be documented, but clearly she did, and the fact that she longed to follow in

---

[3] Emmy's brother Fritz may well have read Kovalevskaya's most famous paper on the motion of a special type of top. Some of his early work dealt with gyroscopes, though mainly in connection with real physical problems. For a detailed analysis of Sofia Kovalevskaya's mathematical work, see [Cooke 1984].

her father's footsteps gives us the first key to understanding how such a thing could have ever happened. As a teenager (Fig. 2.1), Emmy already had to overcome many obstacles before she could even begin her studies at the university.[4] Since her formal schooling ended with the tenth grade, she had to concoct a plan that would later enable her to attend university classes as an auditor. Her French teacher, Mathilde Koenig, may well have been a source of inspiration in this respect, since she was one of three women who received permission to audit classes at Erlangen University in 1897, the year Emmy Noether graduated from the local high school for girls. In all likelihood, Emmy was well aware of this since her father, who was then dean of the philosophical faculty, had written to the Ministry seeking approval for this request. Two years later, Emmy would pass the Bavarian state examination, which qualified her to teach French and English language at schools for young women. She apparently never applied for such a position, however, and probably never intended to do so. She merely took this exam so that she could attend university courses as an auditor, following in the footsteps of her former French teacher. Beyond mathematics, during the next three years she also studied history, romance languages, and archaeology [Tollmien 1990, 160].

What clearly set Emmy Noether apart at this early age was her strong desire to study mathematics rather than pursue a more conventional career, such as to become a schoolteacher. Her family supported this dream by paying for private lessons, though they must have wondered how she would ever be able to take up studies at a university. Only boys were allowed to take the examination required for admission to Bavarian universities, though pressure was mounting to open this exam in special cases to girls, if they were sufficiently prepared. In the meantime, Emmy was allowed to attend courses in Erlangen as a guest, but she also studied alongside her brother Fritz, who was preparing to take his final examinations (*Abitur*) at Erlangen's humanistic Gymnasium after having completed 13 years of schooling there. Fritz passed his exams, whereas his sister had to apply for permission to take a similar test at the Realgymnasium in Nuremberg. Since she had never set foot in this school, she asked for permission to attend classes one week before the exam took place in July 1903. The school director would have approved her request, but the Ministry refused to go along. As it turned out, all went well, but even after she passed this qualifying exam, it was by no means certain that Emmy Noether would be allowed to take up formal studies in Bavaria. Then, two months later, Prince Regent Luitpold signed a decree that gave women the right to matriculate at Bavarian universities if they were properly certified, a true stroke of good luck.

Despite this happy turn of events, Emmy decided instead to begin her studies at the Prussian university in Göttingen, possibly on the advice of her father. This experiment went badly for her, however, and she returned home after only one semester. In the winter semester of 1904, she enrolled at Erlangen University, where only one other female student attended at this time. Although little

---

[4]The following is based on information in [Tollmien 2016a].

Figure 2.1: Earliest known Picture of Emmy Noether, ca. 1900 (Courtesy of MFO, Oberwolfach Research Institute for Mathematics)

is known about the content of the courses she took over the next four years, we may assume she received a solid grounding in algebra and analysis from her father and Gordan. In December 1907, she took her doctorate summa cum laude with a thesis "On the Formation of Systems of Ternary Biquadratic Forms." In it, she followed the classical approach to invariant theory long cultivated by her dissertation advisor, Paul Gordan. This work thus showed no signs of the outstanding theoretician she would become, though it clearly did reveal that she was anything but weak when it came to working through complicated calculations.

After completing her doctorate, Emmy Noether continued her studies while supporting Gordan's and her father's teaching in Erlangen, work for which she received no remuneration. Even though she was already unofficially supervising doctoral students, she had no status beyond her doctoral title, nor did her name appear in the course offerings from this period. Although a small number of postdoc positions were available, these were reserved for men who hoped to climb the career ladder, starting with its lowest rung: an appointment as an unpaid *Privatdozent* (private lecturer). Those who were granted this status by a university faculty had the right (and duty!) to teach, the *venia legendi*, a title conferred after a candidate submitted a second thesis as part of the *Habilitation*. The upshot for Emmy Noether, however, was that, as a woman, she was formally prohibited from applying to habilitate, despite the fact that she was eminently qualified for a position as a *Privatdozent* (see Chapter 4). She had to wait four long years before this prohibition was finally lifted in 1919, the very year in which women around the world were first gaining the right to vote.

Up until the time of Gordan's death in December 1912, she continued to work on invariant theory, publishing two major papers in *Crelles Journal* before taking up other topics in algebra. One can easily imagine that Emmy's father urged her to submit these papers to Kurt Hensel, the editor of this prestigious journal, as a way to promote her work outside the mathematical circles in which Max Noether himself normally moved. Neither he nor Gordan published more than occasional writings in *Crelles Journal*, which was long dominated by the mathematicians from Berlin. Throughout their careers, these two Erlangen mathematicians remained loyal supporters of the rival journal, *Mathematische Annalen*, co-founded by Alfred Clebsch in 1868. In later years, Emmy Noether would follow in their footsteps by publishing most of her lengthier papers in the *Annalen*.

Paul Gordan and Max Noether had been two of Clebsch's closest associates, though otherwise they had little in common. Gordan was voluble and opinionated with a fondness for local cafés and pubs. He loved doing mathematics, which for him meant solving complicated technical problems. What he did not enjoy, however, was the task of writing up his solutions in an intelligible form. Writing was not his forte, and he was happy to leave that task to others. His younger colleague, on the other hand, was one of the leading scholars among mathematicians of his generation. Indeed, Max Noether's writings were by no means restricted to research publications, as can be seen from the many scientific obituaries he wrote for the *Annalen*. Like her father, Emmy Noether was a gifted mathematical expositor, and as a mathematician she had far more in common with him than with her mentor, Paul Gordan.

An anecdote relating to all three took place in early 1913, not long after Gordan's death.[5] In preparing to write an obituary of the latter, Max Noether decided he should pay a visit to his friend Felix Klein in Göttingen. Klein and Gordan had been close collaborators in the 1870s and had remained in fairly close contact for many years afterward. Noether also had the foresight to bring along his daughter; she had been one of only two students who had written a dissertation under the cantankerous Gordan. The discussions between these three were no doubt far-ranging, but also consequential for Emmy, who greatly impressed Klein on this occasion. He and others in Göttingen long remembered how badly she had fared ten years earlier, but on this occasion she struck up a conversation with Klein about his famous book on the icosahedron and fifth-degree equations. By the time it ended, Klein came away feeling like she knew his own book better than he did. Without any doubt, her abilities left a lasting impression on him.

---

[5]Klein recalled this incident more than two years later; see Section 4.2 for other details.

Although he had already formally retired in 1913, Felix Klein continued to preside over mathematical affairs in Göttingen, much as he had for decades. He thismight easily have been forgotten this encounter with Emmy Noether, but when war broke out in August 1914, Klein and his colleague David Hilbert quickly realized that excellent young mathematicians would soon be in short supply. Most of their assistants stood ready to be called into war service, if they had not already volunteered. For the first and probably only time in her life, Emmy Noether could take advantage of the fact that she was a woman. In the spring of 1915, Hilbert and Klein invited her to Göttingen to assist them with their research work, an invitation that carried with it the possibility of *Habilitation*. Of course everyone knew that this was a bold idea; for if this plan were to succeed, Emmy Noether would have become the first female faculty member at a German university.

She applied, in fact, that very year with strong support from the mathematicians on the philosophical faculty in Göttingen. Her supporters, including Klein and Hilbert, tried to make the argument that Noether was an exceptionally talented individual, who could not be compared with other scientifically trained women. They did their best to emphasize that their appeal had nothing to do with creating a precedent that might open the way for other women to pursue academic careers. Instead, they merely requested that the Prussian Ministry of Culture treat this as a special case by granting a one-time exemption from the decree that barred women from becoming *Privatdozenten*. As it turned out, not even the backing of men as powerful as Klein and Hilbert was sufficient to bring this about [Tollmien 1990].

After the ministry rejected this proposal, Hilbert (Fig. 2.4) met with officials in Berlin and successfully negotiated a compromise. Beginning in the winter semester of 1916–17, he offered courses taught by "Prof. Hilbert with the support of Frl. Dr. Noether." In fact, these courses, which mainly dealt with advanced algebraic topics, were taught by Emmy Noether alone. In 1917, the Göttingen mathematicians and physicists contacted the Ministry once again. Their concern was that Noether might leave Göttingen to habilitate at the newly founded university in Frankfurt. But the Ministry responded promptly, assuring her supporters that there was no need for alarm: there was no possibility that Noether, or any other woman, would be allowed to teach at a university. Albert Einstein was also well aware of the circumstances that thwarted Noether's candidacy in Göttingen. Like Klein and Hilbert, he also clearly appreciated Noether's talents. After reading her paper [Noether 1918a], he wrote Hilbert: "It impresses me that one can view these things from such a general standpoint. It would have done no harm to the troops returning to Göttingen from the field if they had been sent to school under Frl. Noether. She appears to know her métier well!" [Siegmund-Schultze 2011a].

Einstein stood in contact with Emmy Noether since 1916, at which time she was working closely with Hilbert on various problems connected with the theory of relativity. One of the most challenging of these problems concerned the status of conservation laws in special and general relativity, a topic of intense discussions in Göttingen. In correspondence with Einstein, Klein emphasized that the usual understanding of energy-momentum conservation in both classical mechanics and special relativity should not be conflated with certain formulations in general relativity derived from variational methods. Hilbert agreed with this, but went further by claiming that this was a characteristic feature of general relativity. He even speculated that one could prove a theorem in support of this distinction. This conjecture was raised in January 1918 when Noether was working with Klein, who wanted to unravel various mathematical difficulties related to energy conservation in general relativity. In response to Hilbert, Klein wrote: "It would interest me very much to see the mathematical proof carried out that you alluded to in your answer" [Klein 1921–23, I: 565]. For Klein, this meant spelling out what it meant to say that a conservation law had real physical content, as opposed to expressing a purely formal property. This, then, became the central problem Emmy Noether set out to solve in early 1918 (Fig. 2.2) in her work on invariant variational problems. Not long afterward, Noether published her now famous paper "Invariante Variationsprobleme," [Noether 1918b] in which she distinguished between different types of conservation laws in theoretical physics.[6]

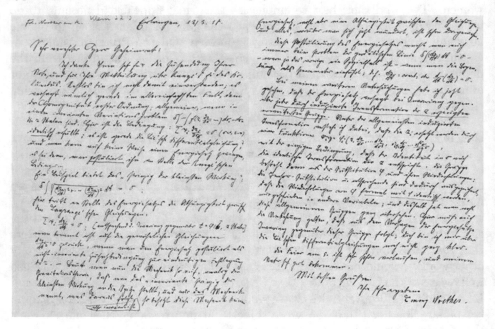

Figure 2.2: Emmy Noether to Felix Klein, 12 March 1918 (Nachlass Klein 22B, SUB Göttingen)

One year later, in June 1919, Emmy Noether submitted this work as her post-doctoral thesis when she was finally allowed to habilitate. Noether's long ordeal in attempting to habilitate constitutes an important chapter in the his-

---

[6]For details, see Chapter 3 of *Emmy Noether: Mathematician Extraordinaire* [Rowe 2020b].

tory of women's struggles for the right to teach at German universities.[7] Emmy Noether's original application to habilitate in Göttingen in 1915 had raised the general issue of whether women might be allowed to assume the duties of a private lecturer at a German university. This question, rather than her self-evident qualifications, stood at the heart of her case, which sparked a sharp debate within the philosophical faculty. Indeed, the Noether affair was perhaps the most infamous in a series of running battles, which would eventually lead to a complete cessation of relations between its two departments, comprised of scientists and humanists, respectively. In 1922, the Ministry approved the proposal forwarded by the historical-philological department that called for the formation of two wholly distinct faculties. The newly established faculty of mathematics and natural sciences then appointed Noether as an honorary associate professor, a title normally awarded only six years after habilitation. In recommending her for this honor, the faculty noted that she had been unjustly denied the right to habilitate in 1915.[8]

During these years, up until her father's death on 13 December 1921, Emmy often left Göttingen to care for him in Erlangen. Max Noether's wife Ida, although eight years younger than he, died six years before him on 9 May 1915. The year before his passing, Emmy officially left the city's Jewish community; her family's religious orientation had probably never been strong.[9] As the oldest of Max Noether's children, Emmy took responsibility for running household affairs. She and her father were not only personally close; Emmy also developed a deep appreciation for Max Noether's place in the mathematics of his time.

As a Privatdozent in Göttingen, Emmy Noether began offering her own courses, which gave her the opportunity to pursue her special mathematical interests while exploring new methodological approaches. Throughout the early 1920s, her principal field of interest was abstract ideal theory. By this time, she had turned away from the Gordanian tradition and instead had adopted a highly abstract standpoint based on axiomatics and algebraic structures. Noether's lectures were considered extremely difficult to follow, both because of the high degree of abstraction but also due to her improvised explanations. Those who preferred polished performances usually avoided her courses, leaving behind a core of devoted listeners who not only grasped what she was saying but who soon took up questions she raised in their own research.

Emmy Noether's passion for "talking mathematics" – long a favorite pastime in the intense mathematical atmosphere that emerged in Göttingen after 1900 – was simply boundless. One of her students, Heinrich Kapferer, recalled a walking tour with Noether, very likely to her favorite spot in the midst of the woods outside town, the *Kerstlingeröder Feld* (Fig. 2.5):

---

[7]The historian Cordula Tollmien gives a detailed account of this story in [Tollmien 1990], which serves as our principal source for Chapter 4.

[8]Universitätsarchiv Göttingen, Personalakte Emmy Noether, UAG.Math.Nat.Pers.9.

[9]According to an article written by Ilse Sponsel to commemorate the fiftieth anniversary of Emmy Noether's death, she resigned from the Erlangen Jewish community on 29 December 1920 (*Erlanger Tageblatt*, 12 April 1985).

It was a beautiful summer's day, but not a word was spoken about the gorgeous surroundings that we encountered, instead it was one on-going mathematical conversation without interruption that lasted for at least two hours. In essence, it was a monologue delivered by E.N. with touching efforts to gain my interest in her problems, but without any kind of written material and with no opportunity for me to take notes that I could refer to later. As such, this was for both of us a strenuous undertaking, for E.N. in a productive sense and for me in a receptive sense, constantly nodding my head out of politeness, even though by no means everything had been fully understood. Finally we came to a clearing flooded with light in the woods , which offered us a well-earned resting spot, no bench, only a soft, rising grassy area, where we could lie down.[10]

Yet not only students found it difficult trying to keep up with Emmy's rapid-fire thought processes. Even Emil Artin, one of the giant figures in twentieth-century mathematics, often found it very difficult to keep up with her. His former wife, Natascha Artin Brunswick, once described how he devised a method for coping with this problem:

They would go for walks, and he would ask her a question, and she would talk *very*, very fast. He knew he couldn't keep up with her, so he would let her talk for about half an hour and then say "Emmy, but I didn't understand a word; could you please tell me again." But in the meantime they would walk very fast, and she would get a little slower and go through it a little more slowly again. The second time he would say, "Emmy I haven't understood it yet." On the third rendition he would understand what she was talking about. By that time, you see, she was so tired that her speed would slow down. She was so amazingly lively! [Kimberling 1981, 34]

Noether's position as "unofficial associate professor" carried no salary, which meant she had to live virtually "hand to mouth" during this period of rampant inflation. Her only regular income came from the modest payments she received for teaching algebra courses, which were paid on the basis of half-year contracts. Since she had no realistic opportunity for professional advancement, Emmy Noether's financial and institutional situation remained somewhat precarious all her life. Although only a lecturer (Fig. 2.3), she supervised a good dozen doctoral dissertations during her Göttingen years. During the winter of 1928/29, she taught as a visiting professor in Moscow, and in the summer of 1930 she substituted for Carl Ludwig Siegel in Frankfurt. One of the many students who heard her lecture in Moscow was Alexandrov's student, Lev Pontryagin, who was impressed by the clarity of Noether's presentations [Koreuber 2015, 172]. Soon afterward, Pontryagin published pioneering work in algebraic and differential topology.

---

[10]Translated from [Tollmien 2016b, 190–191].

Figure 2.3: Emmy Noether at the Podium (Archives of the Mathematisches Forschungsinstitut Oberwolfach)

Despite her lack of professional standing, by the mid-1920s Noether had become a central figure in the social life of Göttingen's vibrant mathematical community. Beyond her seminars and lectures, she interacted with nearly everyone in the institute, including its many guest visitors. In this milieu, Noether became famous for her ambulatory lifestyle, spent ceaselessly "talking mathematics" with her colleagues and students. From the mid-1920s onward, she attracted a number of post-docs, including a few already established mathematicians. Many who attended her courses in Göttingen came away inspired by her approach to modern algebraic concepts and methods. Not a few of these post-docs were foreigners, and in some cases Noether's views and methodology exerted a lasting influence on their research. Together with her doctoral students, they formed what came to be known as the "Noether School," a tightly knit group of mathematicians who soaked up Emmy's enthusiasm for abstract mathematical thinking.

It was soon to be shattered once the Nazi Party came to power in January 1933. Adolf Hitler's movement had long before drawn strong support in Göttingen, in particular within its university's student body. One of the ringleaders among the younger mathematicians was Emmy Noether's former student, Werner Weber, who felt especially emboldened after April 1933, when the Nazi government passed a new Law for the Restoration of the Professional Civil Service [Schappacher 1998, 527–532]. The word "restoration" (Wiederherstellung) was, of course, a euphemism for "purging"; indeed, this law represented only the first of several such measures introduced by the new Nazi regime. It contained an array of criteria, racial as well as political, and was directed not only against Jews but also against anyone whose political views might be regarded as suspect, which included those who were merely critical of the then fervent German nationalist ideology. Supported by Werner Weber and other radicals in the student body, the law was implemented to full effect in Göttingen with especially devastating consequences for mathematics and physics.

Although Emmy Noether had never held a regular professorship, and hence did not belong to the civil service, she was nevertheless swept up in the movement to rid the university of undesirable individuals [Siegmund-Schultze 2009, 60–71]. Soon after the enactment of this "purification law," she and several other Jewish colleagues were informed that they had been placed on temporary leave of absence. Although she had not participated in politics since the early years of the Weimar Republic, conservative elements in the university community considered her to be a radical leftist, a view shared by the *Kurator* of Göttingen University, Theodor Valentiner, who informed the Prussian Ministry of Culture that Noether was, in his view, politically unreliable due to her sympathy for Marxist politics [Tollmien 1990, 206].

These events marked the beginning of a massive exodus of German scholars, who scrambled to find academic positions abroad [Siegmund-Schultze 2009]. In response to this, the Emergency Committee in Aid of Displaced Foreign Scholars was formed in New York City. Two highly prominent figures, Einstein and Weyl, gained appointments at the newly founded Institute for Advanced Study in Princeton. Most, however, had to take whatever work could be found in the middle of the Great Depression.

Emmy Noether was officially discharged in mid-September 1933, after which she lost the small payments she received as a lecturer. Under these circumstances, she had to act quickly. She had hoped to spend part of the coming academic year at Somerville College for women at the University of Oxford, after which she planned to take a position at Bryn Mawr College outside Philadelphia. The Rockefeller Foundation was involved in the financial arrangements at the latter institution, and once these were settled Noether decided to abandon her original plan and move to the United States. In the meantime, her brother Fritz had been dismissed from his post at the Breslau Institute of Technology. Emmy held out hopes that he, too, might find work in America, but her efforts on his behalf proved unsuccessful (see 9.3).

The initiative to create a temporary position for Noether at Bryn Mawr came from Princeton's Solomon Lefschetz, who had become personally acquainted with

Noether during a visit in Göttingen. Lefschetz contacted Anna Pell Wheeler, chair of the mathematics department at Bryn Mawr, suggesting that her college should invite Noether with funding from the Emergency Committee. Bryn Mawr was able to sweeten the deal with additional funds from the Rockefeller Foundation, and in August 1933 offered Noether a salary of $4,000 (worth about $76,000 today) for the coming academic year. Her duties would be to work with advanced students while carrying on with her own research.

Emmy Noether had been forewarned about low academic standards at American colleges for women, so when she arrived at Bryn Mawr in early November 1933 she came with no great expectations. It took her little time, though, to feel quite at home and to take advantage of the opportunities that awaited her. Noether deeply appreciated the recognition she received and the efforts that the college made to integrate her into its community. Beyond the generous financial terms of her contract, Bryn Mawr also established fellowships in Noether's name, thereby aiming to attract young women who wished to pursue careers as professional mathematicians. She struck up a warm friendship with Anna Pell Wheeler, a widowed mathematician who also happened to speak German. Wheeler was Emmy's age and had spent a year in Göttingen studying under Hilbert and Hermann Minkowski before returning to take her doctorate at the University of Chicago in 1910.

Beginning in February 1934, Noether made weekly trips to Princeton, where she presented lectures on modern algebra at the Institute for Advanced Study. Oswald Veblen, a member of the IAS faculty, made arrangements for her talks, which were attended by several eminent mathematicians, including the young algebraist Nathan Jacobson. Many years later, he honored her memory by editing Emmy Noether's *Collected Papers* [Noether 1983].

Noether was quite impressed by the diligence of the four young women who came to Bryn Mawr on fellowships. Three of them – Marie Weiss, Grace Shover, and Olga Taussky – already had completed their doctoral studies, whereas Ruth Caroline Stauffer went on to become Noether's last doctoral student. Just as in Göttingen, where Emmy was nearly always accompanied by at least one of her "Noether boys," she was proud of her "girls," and took them along on her weekly journeys to Princeton and the biweekly graduate mathematics seminar held at the University of Pennsylvania. She already knew the Austrian number theorist, Olga Taussky, from their days together in Göttingen (see Section 7.4). Soon after Noether's death, Taussky moved to England, where she became a Fellow at Girton College, Cambridge University. Three years later, she married the Irish mathematician Jack Todd, with whom she immigrated to the United States after the war. Although she published over 300 research papers, Olga Taussky-Todd taught for many years at the California Institute of Technology before she was finally appointed to a professorship [Goodstein 2020].

The initial financing for Noether's position was scheduled to end in the summer of 1935, but Wheeler hoped that arrangements could be made with the IAS that would enable her to continue teaching at Bryn Mawr. She contacted Veblen, who took up negotiations with the Rockefeller Foundation's Warren Weaver. After some complicated deliberations, Weaver was able to report that the foundation was willing in Noether's case to make an exception to its policy on academic refugees "on the grounds of her unusual eminence" [Shen 2019, 62]. Thus a plan had begun to fall in place to renew Noether's appointment at Bryn Mawr for another two years starting in the academic year 1935/36.

No one realized at the time, however, that Emmy Noether was seriously ill. She underwent an operation to remove a large ovarian cyst, and for three days afterward seemed to be recovering without problems. But on the fourth day she fell into a coma and her temperature shot up to 108 degrees following the rupture of a cerebral blood vessel. Her death on that Sunday, April 14, 1935, came as a complete shock to everyone who knew her as a primordial spirit bursting with energy.

2.2   Modern Algebra and the Noether School

Over the course of her lifetime, Emmy Noether had to overcome many obstacles before she eventually became a world-class research mathematician. Even though she was exposed to higher mathematics from a young age, her true interests only evolved slowly before finally taking on clear form. Her mentor, Paul Gordan, was an old-fashioned invariant theorist; from him she learned special techniques – so-called symbolic methods – for calculating invariants and covariants. As Max Noether wrote in his obituary, Gordan was simply an algorist, and in her doctoral thesis, submitted in 1907, Emmy Noether demonstrated her mastery of the very same approach. No doubt she learned much else, not only from her father but also by studying the mathematical literature; nevertheless, her orientation at this time was grounded exclusively in classical pure mathematics, in particular the study and classification of algebraic forms.

Noether's initial exposure to modern algebra began in 1911 with the arrival of Ernst Fischer as Gordan's successor. Fischer had studied in Vienna, where one of his teachers, Franz Mertens, also happened to be an expert on invariant theory. During his studies, Fischer had also become aware of Hilbert's groundbreaking methods in algebra, which completely overturned the conventional computational techniques employed by Gordan and his generation of algebraists. Much still remained to be done, however, especially since Hilbert had long ago abandoned invariant theory in favor of other fields of research. When Fischer arrived in Erlangen, he found a talented young woman who shared his enthusiasm for the more conceptual approach to algebraic problems. Soon they were spending many hours together, walking, talking and even writing each other. Regretfully, only scant information from their letters and cards has survived, but Noether afterward consistently referred to Ernst Fischer as her true mentor during the period 1911 to 1915. Her published papers from this period abound with references to his work and her many conversations with him. This research continued to focus on invariant theory and related algebraic problems, but through her collaboration with Fischer she gradually came to regard this terrain from a higher plane, a standpoint that at once reflected greater generality combined with conceptual clarity.

A similar urge to generalize can also be seen in her most famous paper from the war years, "Invariant Variational Problems" [Noether 1918b], a work that today is seen as marking a milestone for theoretical physics. Yet as Yvette Kosman-Schwarzbach has shown in [Kosmann-Schwarzbach 2006/2011], it took physicists a very long time to appreciate the significance of what she had accomplished. Noether's contribution arose from conceptual problems in Einstein's general theory of relativity. Some of these critical issues had already surfaced in 1916, when she was working closely with David Hilbert. Two years later, she presented her findings to members of the Göttingen Mathematical Society. Her approach led to extremely general theorems based on a variational system that remains invariant under the infinitesimal generators of a group.

By the early 1920s, Noether's research interests took a decidedly new turn. She herself considered the paper [Noether/Schmeidler 1920], coauthored with Werner Schmeidler, to be the first sign of this new orientation.[11] In 1917 Schmeidler took his doctorate in Göttingen under Edmund Landau, around which time he began working closely with Noether. Under her guidance, he completed his Habilitation in 1919. In their joint paper, they developed a theory of modules for polynomials with a non-commutative multiplication operation, thereby introducing left and right modules, later a central concept in modern algebra [Bourbaki 1960].

In 1921, Noether published her famous paper "Idealtheorie in Ringbereichen" (Ideal Theory in Ring Domains) [Noether 1921b], in which she launched the *general* theory of commutative rings. In it, she dealt with rings whose ideals satisfy the ascending chain condition (acc), which states that every such chain is of finite length. In her honor, the distinguished French mathematician Claude Chevalley later dubbed such rings satisfying the acc *Noetherian rings*. In [Noether 1921b] she famously exploited this general concept to prove what came to be called the Lasker-Noether Theorem, which significantly extended Emanuel Lasker's theorem in [Lasker 1905]. Noether's methods also enabled her to prove a generalized form of Hilbert's finite basis theorem along the way to proving the Lasker-Noether Theorem. Noether's approach eventually spawned modern commutative algebra, the branch of abstract algebra that studies commutative rings, their ideals, and modules over such rings. Modern algebraic geometry and algebraic number theory both developed by drawing on key results in commutative algebra. As Olaf Neumann aptly stated, "it is no exaggeration to say that Noether's 1921 paper was a work of genius which showed the amazing consequences of the ascending chain condition for ideals" [Neumann 2007, 99].

Classical ideal theory was the creation of Richard Dedekind, who introduced new algebraic concepts to gain a deeper understanding of problems in number theory. Over time, Dedekind's works exerted an ever-stronger influence on Emmy

---

[11]This according to [Alexandroff 1935, 2].

Noether's orientation as an algebraist, and she recommended his Supplement XI to Dirichlet's lectures on number theory [Dedekind 1894a] to her students. Nevertheless, her own work was decidedly more abstract than Dedekind's [Corry 2017]. Moreover, her axiomatic approach to algebraic structures opened up new vistas not only for algebra but for other fields as well. The contrast between Dedekind's approach to ideal theory and Noether's was emphasized by Pavel Alexandrov, who first attended her lectures during the summer semester of 1923:

> Of all the lectures I heard in Göttingen that summer, the apex were Emmy Noether's lectures on general ideal theory. As is well known, foundations of this theory had been laid by Dedekind in his famous paper that was published as the eleventh supplement to the edition of Dirichlet's lectures on number theory under Dedekind's editorship. . . . Emmy Noether always said that the whole theory of ideals could already be found in Dedekind and that all she had done was to develop Dedekind's ideas. Of course, the basis of the theory was laid by Dedekind, but only the basis: ideal theory, with all the richness of its ideas and facts, the theory that has exerted such an enormous influence on modern mathematics, was the creation of Emmy Noether. I can judge this, because I know both Dedekind's work, and the fundamental work of Emmy Noether on ideal theory. [Alexandrov 1979/1980, 299]

Throughout the 1920s, Noether's ideas met with considerable skepticism. Some of her colleagues criticized the general trend toward modern axiomatics as overly abstract, and they expressed the fear that these kinds of investigations lacked substance. Similar concerns had been voiced decades earlier in the wake of Georg Cantor's publications on transfinite numbers, ideas that were harshly criticized by leading French mathematicians [Ferreirós 2007, 315–316]. Indeed, this skepticism with regard to Cantor's philosophy of the infinite reflected a prevailing opinion in France that German intellectuals suffered from an excessive fondness for metaphysical theorizing. In this regard, it should not be overlooked that Emmy Noether's mathematical interests were deeply rooted in theories that had been cultivated and developed almost exclusively in Germany. Nor was it an accident that her work soon sparked tremendous interest among the younger generation of mathematicians in France, in particular the group that published under the pseudonym Nicolas Bourbaki [Corry 2004, 220–251, 289–337].

In Germany, Helmut Hasse was one of Noether's staunchest supporters as well as an influential voice in support of the fertility of algebraic methods. He spoke about this in 1929 in Prague at the annual meeting of the German Mathematical Society, emphasizing the importance of abstract algebra in nearly every developing branch of mathematics:

> . . . the modern algebraic method is by no means confined to the body of classical algebra, but rather reaches beyond it and penetrates actually all of mathematics. One can everywhere apply its principle of seeking the simplest conceptual foundation for an existing theory, thereby

Figure 2.4: Hilbert's 60th Birthday, 23 January 1922: l. to r. Auguste Minkowski, unidentified, David Hilbert, Ernst Hellinger, Käthe Hilbert, Lily Minkowski, Emma Schoenflies, Ruth Minkowski, Paul Bernays, Hanna Schoenflies, Peter Debye, Debye, Franz Hilbert, Ferdinand Springer, Otto Blumenthal (Archives of the Mathematisches Forschungsinstitut Oberwolfach)

activating its unifying and systematizing effect, beginning with logic, where one has long spoken of an algebra of logic, and from set theory and foundations, number theory, synthetic and analytic geometry, topology, integral equations, and variational calculus to quantum theory, which has recently entered algebraic waters through the theory of infinite matrices and abstract operators. [Hasse 1930, 33–34]

Emmy Noether was well aware that her style of mathematics represented a break with the past, even if she was fond of citing the work of Dedekind and Ernst Steinitz as forerunners of abstract algebra. She also drew inspiration from classical invariant theory, a field that raised fundamental questions about finiteness, problems that dominated the attention of Noether's teacher, Paul Gordan, as well as the young David Hilbert. Both were problem solvers, who owed much of their fame to the way they cracked open difficult problems pertaining to invariants generated by a finite systems of basis elements. Emmy Noether's fame had little to do with her accomplishments as a problem solver; she was a theoretician with a visionary outlook, a thinker who (to recall van der Waerden's words) strove for "conceptual penetration of her subject matter in order to achieve utmost methodological clarity."

For many famous mathematicians, one can identify a single original idea that served as the driving force behind all their creative work. In Noether's case, there were two: the first dominated during the first two phases of her research career, the second during her third period. Through her expertise in invariant theory, Noether came to recognize the centrality of *finiteness conditions* for abstract algebraic structures, in particular those that came to be called Noetherian. The role of chain conditions in abstract algebra begins with her now classic paper [Noether 1921b] and culminates with the seminal study, "Abstract Construction of Ideal Theory" [Noether 1927].

Emmy Noether's second visionary idea arose through a program to generalize the classical theory of algebraic number fields by appealing to the properties of noncommutative algebraic systems, those in which the commutative law $ab = ba$ no longer always holds. She called this the "principle of application of the noncommutative to the commutative," an idea she promoted in the theory of algebraic number fields and their associated Galois groups. Her long-term aim was to extend class field theory to number fields with noncommutative Galois groups [Curtis 2007]. Perhaps her clearest presentation of this idea came when she was invited to deliver a plenary lecture at the 1932 International Congress of Mathematicians held in Zurich, on which occasion she spoke about "Hypercomplex Systems in their Relationships to Commutative Algebra and Number Theory" [Noether 1932b]. According to her protégé, Pavel Alexandrov, this lecture marked a major triumph for Emmy Noether and the structural approach to mathematics. During this time, she was working closely with Hasse and Richard Brauer; their publications would prove to be of major importance for subsequent developments in modern algebra and algebraic number theory.

Hermann Weyl, who first got to know Noether during the winter semester of 1926/27 when he was a guest professor in Göttingen, later recalled lively conversations with her and John von Neumann following his lectures on the representation theory of continuous groups. Weyl and von Neumann were both then deeply immersed in developing aspects of this theory that were central for the mathematical foundations of quantum mechanics. In his memorial lecture for Noether, Weyl recalled those days:

> I have a vivid recollection of her when I was in Göttingen ... and lectured on representations of continuous groups. She was in the audience; for just at that time the hypercomplex number systems and their representations had caught her interest and I remember many discussions when I walked home after the lectures, with her and von Neumann, who was in Göttingen as a Rockefeller Fellow, through the cold, dirty, rain-wet streets of Göttingen. [Weyl 1968, 3: 432]

Emmy Noether was not a particularly prolific mathematician. Her *Collected Papers* were compiled in a single volume in [Noether 1983], and while a few of her papers would become famous, many have long since been forgotten. Noether's fame and influence had much to with those well-known publications, of course, but one cannot really begin to grasp her importance merely by studying these published works. For this would be to overlook her activities as a collaborator and critic, not to mention her role as a referee for the journal *Mathematische Annalen*. Noether lived during the pre-Bourbaki era, a time when modern forms of collaboration were only emerging. André Weil, the unofficial leader of the group that wrote under the pseudonym Nicolas Bourbaki, remembered the atmosphere in Noether's Göttingen circle during the mid-1920s as very different from the one he encountered when talking with those in Courant's group, from whom he learned very little. Nearly every time he got into a conversation with one of the latter's students, the exchange would end rather abruptly with a remark like, "sorry, I have to go write a chapter for Courant's book" [Weil 1992, 51]. This "publish or perish" mentality predominated in Courant's circle, whereas Emmy Noether felt no such urgency to rush her work into print. Weil recalled conversations with Pavel Alexandrov in Noether's cramped little attic apartment (Fig. 6.3). Its ceiling was so angular that Edmund Landau – who lived in a veritable palace by comparison – wondered out loud whether Euler's polyhedral formula still applied to her living room. Weil later recalled this easygoing atmosphere, and how Emmy Noether

> good-naturedly played the role of mother hen and guardian angel, constantly clucking away in the midst of a group from which van der Waerden and Grell stood out. Her courses would have been more useful had they been less chaotic, but nevertheless it was in this setting, and in conversations with her entourage, that I was initiated into what was beginning to be called "modern algebra" and, more specifically, into the theory of ideals in polynomial rings. [Weil 1992, 51]

Many who heard Noether's lectures reacted similarly, like the young Carl Ludwig Siegel, who remembered her lectures as badly prepared. In one of them, which ended at 1 o'clock, he scribbled in the margin of a notebook: "It's 12:50, thank God!" [Dick 1970/1981, 1981: 37]. Siegel much preferred lectures in the style of his mentor Landau, who prided himself on presenting polished lectures already nearly ripe for publication. Noether, on the other hand, developed a teaching style that favored dialogue and collaboration. Doing mathematics meant, for her, engaging with all facets of the process, and in this way she came to embody the oral component of Göttingen's vibrant mathematical culture.[12] This approach should, however, by no means be understood as neglecting the importance of formal rigor and written forms of communication. Indeed, the relative sparsity of her own published work reflects the fact that she always resisted rushing her latest work into print. Moreover, Noether's letters and postcards – in particular those

---

[12]On the importance of the oral dimension in Göttingen, see [Rowe 2004].

she sent to Helmut Hasse, published in [Lemmermeyer/Roquette 2006] – reveal very clearly that she upheld the highest standards for mathematical publications. "Pauca sed matura" (few but ripe), the famous watchword of Carl Friedrich Gauss, applies just as well to Emmy Noether. Yet Gauss, who was anything but generous when it came to communicating his unripe ideas with others, stands in this respect in complete opposition to Emmy Noether, whose success and influence had much to do with her unselfish generosity. Indeed, despite her unorthodox teaching style, she made her greatest impact as a teacher and as the leader of what came to be called the "Noether school" (or "Noether community" or sometimes "Noether family").

The Noether school formed an open-ended intellectual space, one in which Emmy Noether challenged younger mathematicians to learn how to exploit the new concepts and methods she promoted in her lectures and published work. Her approach aimed to strip mathematical objects down to their bare essentials in order to recognize deeper underlying relationships among them. Doing so, however, meant learning to think about mathematics on a higher abstract plane. Although her methodological principles were algebraic, they found fertile applications in other fields, including number theory, geometry, topology, and analysis. Indeed, Emmy Noether saw herself as riding a wave of modern methods that were creating a new mathematics founded on abstract axiomatics and algebraic structures. Her enthusiasm was infectious, and those who entered her circle soon realized that she was at the cutting edge of a major movement in mathematics under the watchword of abstract algebra. Soon several members of this Noether school were making their own contributions to this new modern algebra.

Mathematical schools have had a long and venerable tradition in Germany, though the "Noether school" in some ways represents a break with the past. In general, scientific schools tended to foster close relationships between professors and students. Even today, mathematicians often use familial language to suggest a certain mixture of personal and intellectual intimacy that professors have traditionally cultivated with their students. Emmy Noether, who never held a regular university professorship, certainly promoted warm relations with her doctoral students (Fig. 2.5). Even as a post-doc in Erlangen, where she had no official position whatsoever, she served as adviser to two students who did their thesis work under her.[13] Nevertheless, her larger influence extended far beyond the immediate circle of her doctoral students. One of the first to benefit from her support was Landau's student, Werner Schmeidler. Already in early 1920, she took full advantage of her connections within the far-flung Göttingen network to promote his future career at two provincial outposts, Kiel University and the recently founded Breslau Institute of Technology. One should keep in mind that the number of permanent positions in mathematics at this time was very small, numbering no more than

---

[13]For detailed information on the work of Noether's doctoral students, see [Koreuber 2015, 159–195, 310–336].

200 at all institutions of higher education throughout all of the German states. It was a tiny world.

Everyone knew that Emmy Noether took a deep interest in the professional development of all her associates, including, of course, her doctoral candidates. Among her most prominent students were Ernst Witt, Max Deuring, Grete Hermann, Heinrich Grell, Jakob Levitzki, Hans Fitting, and Chiungtze Tsen.[14] Beyond this group, however, there were numerous others who worked closely with Emmy Noether, and who in a broader sense belonged to the "Noether school." The four supporting actors in "Diving into Math with Emmy Noether" – B. L. van der Waerden, Helmut Hasse, Olga Taussky, and Pavel Alexandrov – were all associates of this extended Noether school (see Chapters 5, 6, and 7). Each had a distinctively different background and all four entered her circle as mature mathematicians pursuing independent lines of research. Hasse and Noether worked together closely as collaborators, whereas the other three were influenced by Emmy Noether's ideas and personality in specific ways clearly reflected in their published work.

Emmy Noether's school was also part of a larger network of mathematicians, who in one form or another were promoting modern algebra in the 1920s and early 1930s. An often overlooked figure was Alfred Loewy, who spent his entire career in Freiburg. He habilitated in 1897, then was promoted to an associate professorship in 1902, before gaining a full professorship in 1919. The Nazis abruptly ended his teaching career in 1933, forcing him into early retirement. Loewy's position then went to Wilhelm Süss, who soon emerged as one the most influential mathematicians in Germany.[15] Two close associates of Emmy Noether, Wolfgang Krull and Friedrich Karl Schmidt, began their careers working under Loewy, and three others with close ties to Noether also studied under him in Freiburg: Ernst Witt, Richard Brauer, and Reinhold Baer. Witt later became one of Noether's star students in Göttingen, Brauer was one of her most important collaborators, and Baer became part of the Noether-Hasse network in 1928 when he joined Helmut Hasse as his assistant in Halle. There they published a new edition of Steinitz's classic study on field theory [Steinitz 1930] together with an appendix on modern Galois theory, which appeared in the same year as volume 1 of van der Waerden's *Moderne Algebra* [van der Waerden 1930/31].

During the 1920s, Richard Courant and Emmy Noether actively promoted the trend toward internationalization that became such a striking feature of Göttingen mathematics during this period. Their efforts received a major boost from American philanthropy and the vision of Wickliffe Rose, who founded the International Education Board (IEB) in 1923, backed by financial support from John D. Rockefeller, jun.[16] Several of those who visited Göttingen during these years

---

[14]On the gradual development of Emmy Noether's school and its broad influence, see [Koreuber 2015, 159–195].

[15]On Loewy's career and mathematics in Freiburg, see [Remmert 1995].

[16]For a detailed account of the IEB's impact on mathematics, especially in Western Europe after World War I, see [Siegmund-Schultze 2001].

Figure 2.5: On an *Ausflug* to Kerstlingeröder Feld, l. to r.: Otto Schilling, Emmy Noether, Olga Taussky, Hans Schwerdtfeger, Ernst Witt, Paul Bernays, unidentified, Erna Bannow, unidentified, summer 1932; Papers of John Todd and Olga Taussky-Todd, Caltech Archives.

were IEB Fellows. Two who came from France were André Weil and Paul Dubreil; both attended Noether's lectures, as did another co-founder of the Bourbaki group, Claude Chevalley. The Norwegian Øystein Ore visited Göttingen twice, the second time as an IEB fellow working under Noether. He was afterward recruited by James Pierpont, who invited him to join the faculty at Yale University, where he would remain throughout his career. He also joined Emmy Noether and Robert Fricke in editing the collected works of Richard Dedekind, [Dedekind 1930–32] (see 7.4).

The first official graduate of the "Noether school," Grete Hermann (Fig. 2.6), took her degree in 1926 with an important dissertation on ideal theory [Hermann 1926]. In it she gave the first algorithm for computing primary decompositions of polynomial rings, a method still used today in computer algebra. Other German students of Noether included Max Deuring, Heinrich Grell, and Ernst Witt, who mingled with the foreigners in her circle. One of these was Jakob Levitzki, who took his doctorate under Noether in 1929 before joining the faculty at the Hebrew University in Jerusalem two years later. Another was her doctoral student Chiungtze Tsen (1898–1940), who introduced modern algebra in China.

Figure 2.6: Emmy Noether and Grete Hermann, June 1926 (Auguste Dick Papers, 13-1, Austrian Academy of Sciences, Vienna)

A third was Kenjiro Shoda, a student of Teiji Takagi who came to Germany on a fellowship from the University of Tokyo. Shoda spent his first year abroad studying algebra under Isaai Schur, before arriving in Göttingen in 1928. There he attended Noether's lectures on hypercomplex systems and representation theory, an experience that Hiroshi Nagao considered decisive for his subsequent career:

> This particular year seems to mark the most significant period in his mathematical growth. There, near Noether, he witnessed the remarkable process of creation of great mathematical ideas and theory, and the youthful Shoda buried himself in enthusiastic pursuit of mathematics in a wonderful creative atmosphere generated by the many young, able mathematicians who had come from all over the world to Göttingen, attracted by Emmy Noether. [Koreuber 2015, 169]

Kenjiro Shoda's textbook *Abstract Algebra* from 1932 helped popularize the subject in Japan.

Among the foreign mathematicians associated with the Noether school, the two most prominent were Bartel van der Waerden, who came to Göttingen from Amsterdam in 1924, and Pavel Alexandrov from Moscow, who first visited one year earlier. Both enjoyed very close relations with Emmy Noether, as described in greater detail in Chapters 5 and 6. Since van der Waerden's main research

interests were in algebraic geometry, he had already studied some of the works
of Max Noether before he met Noether's daughter, from whom he first learned
about applications of modern algebra. Alexandrov's situation was quite similar,
except that he happened to be a topologist. He, too, knew nothing about abstract
algebra until he met Emmy Noether. Only shortly after her death, Alexandrov
spoke movingly about the atmosphere that surrounded her in Göttingen:

> Emmy Noether had close ties with Moscow. These ties began in 1923
> when the late Pavel Samuelovich Urysohn and I first arrived in Göttin-
> gen and immediately found ourselves in the mathematical circle led by
> Emmy Noether. We were immediately struck by the fundamental traits
> of the Noether school: the mathematical enthusiasm of its leader, which
> she conveyed to all her students, her deep conviction in the importance
> and the mathematical fertility of her ideas – a conviction which was far
> from universally shared even in Göttingen – and the extraordinary sim-
> plicity and warmth of the relations between Noether and her students.
> At that time, this school consisted almost entirely of young Göttingen
> students. [Alexandroff 1935, 7–8]

During the mid-1920s, Alexandrov began a new collaboration with Heinz
Hopf; both were regular guests in Göttingen. There they enjoyed many inspiring
conversations with Noether, who keenly followed their joint work. Over the period
from 1925 to 1931, they presented their new topological ideas in special lectures
and seminars attended by Emmy Noether and Richard Courant. At some point,
Courant suggested that they should rework this material into a monograph for
Springer's "yellow series." When *Topologie* [Alexandroff/Hopf 1935], a pioneering
text for algebraic topology, finally appeared shortly after Emmy Noether's death,
the authors underscored their debt to her, thereby bearing witness to their own
close affiliation with the Noether school.

Alexandrov's friendship with Emmy Noether intensified during the winter of
1928/29, when she offered a seminar on modern algebra in Moscow. In an undated
letter to Oswald Veblen from 1928, he wrote:

> This winter (as you know) we have Miss Noether here in Moscow as a
> guest professor, and, of course, her presence enlivens our mathematical
> life greatly, especially since algebra notably belongs to those mathe-
> matical fields that unfortunately have been cultivated little so far in
> Moscow. Partly under the influence of Miss Noether, partly also stim-
> ulated by my algebraic lectures here in Smolensk, I begin to become
> very interested in algebra, for the time being only "from afar" of course,
> without trying to work in it myself. The topological lecture which Hopf
> is holding in Berlin this winter (and which is very interesting [...]) is
> also very algebraically influenced. [Merzbach 1983, 168]

Veblen had corresponded with Noether a year earlier in connection with her work
on differential invariants. His curiosity about this was surely piqued by reading

in [Noether 1918a, 44] that she planned to publish a more detailed account in
Mathematische Annalen. As so often happened, however, she never found time
to complete such a paper and had to inform Veblen that her interests were now
focused on arithmetical matters [Merzbach 1983, 165].

2.3 Belated Recognition

Despite the appreciation of mathematicians like Alexandrov, Hopf, and Veblen – whose research fields were far removed from her principal interests – Emmy
Noether may well have felt unappreciated at times, even by those who accepted
her views and followed in her wake. After all, very few mathematicians cited her
work with the same generosity she showed toward others in her own publications.
After Hermann Weyl assumed Hilbert's chair in 1930, he tried to nominate her
for membership in the Göttingen Scientific Society, but without success. At the
time of her death, he also spoke of another injustice she suffered, one for which
Hilbert and his allies were alone to blame. As he rightly noted, "Emmy Noether
was a zealous collaborator in the editing of the *Mathematische Annalen*. That this
work was never explicitly recognized may have caused her some pain" [Weyl 1935,
442–443]. A telling statement she herself made comes from a letter she wrote to
Helmut Hasse on 12 November 1931: "My methods are really methods of working and of thinking; which is why they have crept in everywhere anonymously"
[Koreuber 2015, 157]. In all likelihood, this was not really meant as a complaint
so much as an expression of her satisfaction in seeing the way others had taken up
her ideas.

Around this time, in fact, Noether was finally beginning to receive outward
signs of long overdue recognition. In 1932, she and her Hamburg colleague Emil
Artin were awarded the Ackermann-Teubner Memorial Prize for their achievements in modern algebra. During the eighteenth and nineteenth centuries, mathematical prizes were mainly conferred by scientific academies for special works
submitted as answers to problems set by these institutions [Gray 2006]. Two of
the most famous female mathematicians, Sophie Germain and Sofia Kovalevskaya,
both gained fame by winning prizes offered by the Paris Academy.

In the modern era, but before the establishment of the Fields Medals and
other prizes for distinguished mathematical work, the Ackermann-Teubner Prize
represented the highest award offered in Germany. Created in 1912 by Alfred
Ackermann-Teubner, the Leipzig publisher and longtime benefactor of the German
Mathematical Society, it was first conferred on Felix Klein in 1914 following the
wishes of its founder. Every two years, one of eight areas of research in pure
or applied mathematics became eligible, at which time a five-member jury would
select the winner. In 1932 this area was arithmetic and algebra, two fields in which
Emil Artin had recently made highly significant contributions. In 1927 he solved
two of Hilbert's 23 Paris problems – the 9th (to establish a general reciprocity
law for algebraic number fields) and the 12th (to express a nonnegative rational

function as a quotient of sums of squares) – making him an obvious candidate for the prize.

Martina Schneider recently uncovered some of the actions behind the scenes that led to Noether's selection along with Artin. Two of the five jurors were van der Waerden and Erich Hecke, who was Artin's older colleague in Hamburg. In a letter addressed to Hecke, but sent to the other three jurors as well, van der Waerden argued forcefully in support of Emmy Noether's case:

> ... not only because of her mathematical achievements, but above all because of the extraordinarily *stimulating* and *directive effect* she has exerted on a whole generation of algebraists. It was she who made all of us, including Artin, aware of Steinitz's work, and who in her work on elimination initiated applications of field-theoretical methods and concepts to algebraic geometry. It was she who created general ideal theory, with the chain conditions and the proof of the general decomposition theorem .... It was she who, against all odds, repeatedly emphasized Dedekind's module methods and extended these to group theory and ring theory, until these methods led to her triumph with the theory of ideal and module classes of hypercomplex systems, namely the long-desired unification of hypercomplex numbers and representation theory. Finally, it was she who long ago foresaw and pointed the way to the number-theoretical applications of hypercomplex theory, which Hasse and others are now successfully pursuing. ...
>
> You can already see from the above that I would favor awarding the prize to Noether. There is still another motive that influences me to tend more toward Noether in this difficult comparison of the diverse achievements of Artin and Noether: the consideration that she has not yet received as many honors as Artin, who has already received various offers [for professorships], whereas she is only a private lecturer with a teaching contract. One could try to use this opportunity as a form of compensation.[17]

This turned out to be the only instance when two individuals were chosen as recipients of the Ackermann-Teubner Prize.

The flowering of modern algebra soon led to the publication of major monographs, several of which were written by members of the extended Noether school: Wolfgang Krull, Max Deuring, Shoda, and van der Waerden.[18] Krull's *Idealtheorie* [Krull 1935] provided a compendium of results in this field, which had grown into a major new discipline starting with Noether's works from the early 1920s. Deuring, who took his doctorate under Noether in 1930, was one of her closest students. His monograph *Algebren* [Deuring 1935], like Krull's on ideal theory, was published in Springer's *Ergebnisberichte* in 1935. In it, he summarized recent work

---

[17]Van der Waerden to Hecke, 29 May 1932, translated from [Schneider 2021].

[18]For an overview of these three seminal texts – [Krull 1935], [Deuring 1935], and [van der Waerden 1930/31] – see [Koreuber 2015, 232–269].

on the modernization of the theory of hypercomplex systems. Noether proofread the text during her last visit to Göttingen in the summer of 1934.

The most famous and influential new monograph, however, was B.L. van der Waerden's *Moderne Algebra* [van der Waerden 1930/31], published in two volumes. This was one of the first algebra textbooks to present the subject starting with the axioms for three fundamental structures: groups, rings, and fields. Plans for this book actually dated back to 1926, when van der Waerden attended Emil Artin's lectures in Hamburg. The latter had made arrangements to publish a textbook on modern algebra in Richard Courant's "yellow series" (*Grundlehren der mathematischen Wissenschaften*), a project he originally intended to carry out with the help of the young Dutchman. Many years later in an interview with the second author, van der Waerden gave this version of what then happened:

> [Artin] wanted to write the book with me and so I began writing, finished a chapter and brought it to Artin, and asked him whether he was satisfied with it. He was satisfied, and so I asked if he had begun to write a chapter. He had not, but I kept writing, finished a second chapter and brought it to him. Then he said: "No, you go ahead and write the book." [Koreuber 2015, 235]

This highly contracted, retrospective account seems nevertheless quite believable, especially considering that van der Waerden was a highly prolific writer, whereas Artin preferred the spoken word, much like Emmy Noether. In any event, we find the strikingly different interpretation of these events in [Soifer 2015] entirely unconvincing.[19]

B.L. van der Waerden obviously did not write *Moderne Algebra* in a matter of months. Following his stay in Hamburg, he habilitated in Göttingen, where he began his teaching career in the winter semester 1927/28 by offering a course on ideal theory. That same semester, he also attended Noether's course on hypercomplex numbers and group characters, and together they reworked his notes for publication as [Noether 1929a]. Van der Waerden then combined this material with Artin's lectures, while drawing on various other works. Later, he described this process in detail in [van der Waerden 1975], which gives a clear account of the sources he used in writing [van der Waerden 1930/31]. The revised edition he published in 1970 was entitled simply *Algebra*, since the adjective "modern" no longer seemed relevant. When the ninth edition appeared in 1993, Jürgen Neukirch commented in his Vorwort that van der Waerden's now classic text, "with its new abstract conceptual treatment of algebra was intellectually and temporally a product of the twentieth century and a harbinger of the future. From its subtitle, 'after lectures of E. Noether and E. Artin,' we can indeed sense the highly modern conceptual thinking of Emmy Noether and the elegance of Artin's train of thought."[20]

---

[19]A number of other problematic features are discussed in [Siegmund-Schultze 2015] .

[20]Translated from [Koreuber 2015, 244].

In the meantime, memories of Emmy Noether and her vibrant mathematical school lived on, despite her death and the destruction of the Göttingen mathematical community in which she had thrived. During the diaspora that began when Hitler came to power in 1933, the process of internationalization, long a hallmark of the Göttingen mathematical tradition, gained new momentum. Once Noether's distinctive approach to modern algebra had found its way into print, the import of her work could spread more easily, and as it did her legacy gradually grew. Soon modern algebra took root at leading mathematical centers in the United States and Russia as well as in France and Japan before expanding to other parts of the globe. Van der Waerden, in fact, already predicted this in his obituary, which ended with the words: "today the triumphal march of modern algebra based on her thoughts will be unstoppable all over the world" [van der Waerden 1935, 474].

Chapter 3

Max and Emmy Noether: Mathematics in Erlangen

Until 1933, most of Emmy Noether's life was spent in two middle-sized cities: Erlangen, her birthplace, and Göttingen, where she began her mathematical career. Noether was already thirty-three when she left Erlangen for Göttingen in 1915. Although her brilliant career as an algebraist only began after her habilitation in 1919, one can trace many roots of her later mathematical activity and the work that would later make her famous back to Erlangen. The university's mathematical faculty, one of the smallest in Germany, had only two members. Both also happened to be of Jewish descent: Emmy's doctoral advisor, Paul Gordan, and her father, Max Noether, a leading algebraic geometer. These circumstances were highly unusual, making Erlangen an important locale for gauging the careers of Jewish mathematicians, as will be seen in this chapter.

In Erlangen, but also during her first years in Göttingen, Emmy Noether was primarily known as the daughter of Max Noether. Today he is mainly known as the father of the famous "mother of modern algebra." Aside from this not uninteresting observation, Max and Emmy Noether have seldom been compared, even though there are plenty of indications that she studied her father's works in detail. Moreover, careful examination of her earlier work clearly reveals streams of thought from her Erlangen period that flowed into her later work in Göttingen. Like her father, Emmy was an impressive scholar, a mathematician whose work evinced broad and detailed knowledge of the mathematical literature. In this respect, both were outstanding representatives of Germany's high mathematical culture, to which they made fundamental contributions. Yet Max Noether has rarely received serious attention in the by now quite extensive literature devoted to his daughter Emmy. Not only in this chapter but elsewhere in this book, Max Noether's name comes up often, and this for good reason: he was most definitely a major formative influence on Emmy Noether's life. This chapter thus aims, among

© The Author(s), under exclusive license to Springer Nature Switzerland AG 2020
D. E. Rowe, M. Koreuber, *Proving It Her Way*, https://doi.org/10.1007/978-3-030-62811-6_3

other things, to shed a small beam of light on the relationship between these two great mathematicians, who in several respects belong together.

3.1   Max Noether's early Career

To gain a sense of Emmy's early development, one must go back to her years in Erlangen (Fig. 3.1), beginning with her home life there as the oldest of four children and the only daughter of Max and Ida Noether. Emmy Noether's mother grew up in a large and very wealthy family from Cologne; she was one of eleven children of Markus Kaufmann and his wife Frederike Kaufmann née Scheuer. Two of Ida Kaufmann's brothers assisted Emmy Noether financially after the death of her parents. These were her two uncles in Berlin: Paul, a wholesale merchant, and Wilhelm, a university professor who specialized in international economics [Dick 1970/1981, 1981: 8].

Figure 3.1: Emmy Noether's Place of Birth in Erlangen, Photo from March 1982 (Auguste Dick Papers, 12-14, Austrian Academy of Sciences, Vienna)

After her father's death in 1866, Ida moved with her mother to Wiesbaden, a city known for its spas and aristocratic culture. Up until that year, Wiesbaden had been the capital of the Duchy of Nassau, but having sided with the Austrians in the Austro-Prussian War it fell into the hands of the Hohenzollern monarchy. During the era of the Kaiserreich, the emperors began making annual summer trips to Wiesbaden, which led to a construction boom that continued up until the First World War. It was in this glamorous city in 1880 that Max Noether married Ida Kaufmann, who would spend the remainder of her life looking after their household in Erlangen. Although little is known about Emmy Noether's mother, Auguste Dick reported that she enjoyed playing the piano all her life, a talent she tried to pass on to her daughter, but without success [Dick 1970/1981, 1981: 9–10]. How she and her husband first met is also unknown; since marriages were still quite often arranged during this era, the couple may have barely known one another when they wed.[1] Ida Noether's family no doubt offered a substantial dowry at the time, which surely made life in Erlangen for the young family more comfortable. Max Noether's salary as an associate professor was considerably less than that of a full professor (*Ordinarius*), and he would only gain that coveted title eight years later.

Beginning with the nineteenth century, the city of Erlangen belonged to Bavaria. Its citizenry was fairly equally divided between Catholics and Protestants, whereas Jews were only allowed to settle in the city after 1861. Until then, fairly large Jewish communities existed in outlying villages, where life was hard and poverty widespread. A decade later, after the unification of Germany under the domination of Prussia, 65 Jews were living in Erlangen, a city of some 12,500 inhabitants. That number steadily rose to around 240 in 1890, which was roughly 1.5% of the total population. In the meantime, Jewish life in the outlying villages nearly disappeared as the flight to larger cities took place throughout large parts of Germany.[2] Here real economic opportunities awaited them, and the German Jews contributed greatly to the modernization of urban centers in nearly all parts of the German Empire. When economic misfortune struck, on the other hand, as happened in the early 1870s, the blame often fell on Jewish financiers. This was a new form of antisemitism, a hatred tinged by envy rather than the loathing of Christian society.[3]

To what extent Max and Ida Noether's children were exposed to milder forms of prejudice against Jews no one will likely know. They belonged to a special elite, as the offspring of a university professor, and their parents may well have avoided talking about antisemitism in their presence. Emmy grew up with three younger brothers: Alfred, Fritz, and Gustav Robert. By all reports, she enjoyed a happy childhood, but her mother's life was hardly carefree, as two of her sons had serious health problems. Alfred, the eldest, had a weak constitution and died near the

---

[1] Marion A. Kaplan describes the era of the Kaiserreich as a transitional period for Jewish families, as they began to allow young couples limited freedom in choosing a partner [Kaplan 1991].

[2] A very similar pattern can be seen in the case of Göttingen, where the Jewish population nearly tripled between 1867 and 1885; for a detailed study, see [Wilhelm 1979].

[3] The distinction between modern antisemitism and traditional religious forms was made by Hannah Arendt in the first essay in her study *The Origins of Totalitarianism* (1951).

end of the war at age 35. Robert, the youngest, was mentally handicapped and spent his last years in a sanatorium; he died before reaching the age of 40.

Fritz, on the other hand, was healthy and robust. He and his sister were very close all their lives, though temperamentally they differed quite strikingly. Fritz was more serious and sober-minded, whereas Emmy had a fun-loving spirit. As a professor's daughter, she looked forward to dancing parties at the houses of Max Noether's colleagues [Dick 1970/1981, 1981: 11]. Her easygoing manner no doubt led people to overlook that she was also ambitious and self-disciplined; already as a teenager she knew that she wanted to study mathematics, perhaps even follow in her father's footsteps [Tollmien 2016a]. When she first had such a dream is impossible to say, but since her brother Fritz had similar thoughts, it seems more than likely that both talked about such plans for the future. Moreover, in one sense they were in a privileged situation. How many teenagers could even imagine the kind of life their father led, constantly steeped in thought about matters neither they nor their mother could comprehend? Yet this was a natural part of their home life, and so they grew up knowing instinctively how mathematicians think and talk, and also how they demand peace and quiet to concentrate on their work.

Very few published sources contain information about Max Noether's early life, and those that happen to report on his youth invariably obtained those facts from his daughter.[4] Max Noether (Fig. 3.2) was born in Mannheim on September 24, 1844, as the third of five siblings. His father and an uncle ran a well-established wholesale iron business that provided their families with financial stability. According to Emmy Noether, her father was very close with his mother, although she knew this only through him; Emmy's grandmother died long before she was born. Max was an ambitious child and went straight into the third grade of the Gymnasium after primary school. At the age of 14, however, he contracted polio, which resulted in paralysis in one of his legs. For the next few years he could barely walk at all. He took private lessons during this time, but as Emmy reported, he also spent many hours reading. It was thus through self-initiative that "he laid the foundation for a very extensive literary and historical education. At home he took it upon himself to work through the usual university curriculum in mathematics" [Brill 1923, 212–213]. Under very different circumstances, Emmy Noether would later do the same when preparing to take the examination required for admission to a university.

Max Noether's first love was astronomy, which he pursued at the local Mannheim observatory. His first publication was a short paper on the paths of comets; it appeared in *Astronomische Nachrichten* in 1867 when he was still a student at Heidelberg University. More than twenty years later, having long since made a name for himself in algebraic geometry, he published a lengthy review of Henri Poincaré's famous prize-winning study of the 3-body problem [Barrow-Green 1997]. In Heidelberg, Noether mainly studied theoretical physics under Gustav Kirchhoff. Emmy Noether commented briefly on how Kirchhoff indirectly kindled her father's early

---

[4]This applies to [Brill 1923] as well as for [Castelnuovo/Enriques/Severi 1925].

mathematical interests by way of mapping problems in theoretical physics. These led him to Riemann's works and then to the geometric theory of algebraic functions, which he learned by reading Riemann as well as the monograph by Clebsch and Gordan [Brill 1923, 213]. Noether needed only three semesters to complete his doctorate in Heidelberg. At that time, a dissertation was not even required, but he nevertheless submitted his astronomical work as a doctoral thesis, only to have it returned to him. In the end, Noether merely had to endure an easy "oral examination in the dean's apartment, for which the doctoral student was obligated to supply the wine" [Brill 1923, 213]. These details we owe to Emmy Noether's recollections of her father's early life.

In Heidelberg Max Noether also befriended Jakob Lüroth, who habilitated there after studying under Alfred Clebsch in Giessen. On Lüroth's advice, Noether left for Giessen in 1868, a decision that would decisively influence the course of his subsequent career. Clebsch had invited Paul Gordan to habilitate in Giessen, where he taught as a *Privatdozent* until his promotion to associate professor in 1865. Three years later, Clebsch assumed Riemann's chair in Göttingen, and in 1869 Gordan married Sophie Deurer, the daughter of a professor of law in Giessen. Noether was by now strongly drawn to Clebsch, so he left Giessen to continue working under him in Göttingen. In a letter from Göttingen, written to his future collaborator Alexander Brill on July 7, 1869, Noether soberly noted: "The work I hereby send to you, as you will see, stems from the sphere of Clebsch's findings, though I claim for myself the ideas developed and hinted at therein" [Brill 1923, 214]. Clebsch was much impressed by Noether's new results; he later told Brill, he would have been even happier had he found them himself [Brill 1923, 215]. Around this same time, Felix Klein came to Göttingen from Bonn to study with Clebsch, who was by now the head of a prominent mathematical school [Tobies 2019, 37–48]. One year earlier, Clebsch and Carl Neumann founded the journal *Mathematische Annalen*, which later would become the main publishing organ for mathematicians with close ties to the Göttingen network.

Noether and Klein soon became close friends – adopting the more intimate "du" form when they addressed each other – a friendship they maintained up until Noether's death in 1921. Although their time together in Göttingen was brief, it was also very significant for both of them. Klein left for Berlin in the fall of 1869, and then in the spring of 1870 he went to Paris, where he joined his newfound Norwegian friend Sophus Lie. Their stay, however, ended abruptly, when in mid-July France declared war on Prussia. Klein returned home quickly, joined a crew of emergency volunteers, and returned to France, where he witnessed the battle sites around Metz and Sedan, before falling ill. After spending several weeks recovering from gastric fever at his family's home in Düsseldorf, he habilitated in January 1871 in Göttingen, under the watchful eye of Clebsch. By now, however, Noether was already back in Heidelberg, where he habilitated in the winter semester 1870/71. During all this time, Klein and Noether corresponded regularly, not least because their mutual mathematical interests were very close during these years.

The friendship that developed between Klein and Noether clearly had much to do with the fact that both enjoyed close ties with Clebsch. Three years later, in November 1872, both were deeply shocked when they learned that their revered master, who was not yet 40 years old, had suddenly died from an attack of diphtheria. Only a short time before his death, Clebsch had paved the way for Klein – who was then only 23 years old – to be appointed as the new professor of mathematics in Erlangen. In so doing, Clebsch passed over two far older candidates from his school, namely, Gordan and Noether. Klein remained in Erlangen for only three years, yet his name remains prominently associated with this university owing to his famous Erlangen Program [Klein 1872], which he published in 1872 when he joined its philosophical faculty. He was then its only mathematician, but in 1874, the year before he succeeded Otto Hesse in Munich, Klein managed to gain a second position for Erlangen. He also arranged for Paul Gordan, Emmy Noether's future doctoral supervisor, to fill this associate professorship. This enabled Gordan to assume Klein's chair one year later, thereby opening the door for Max Noether to fill Gordan's post as associate professor. It was a classical case of networking, but with long-term significance, since these arrangements helped to stabilize the precarious state of the Clebsch school and its journal, *Mathematische Annalen*. Klein and Gordan continued to collaborate during the years that followed, often meeting in the small town of Eichstätt, which was conveniently located halfway between Erlangen and Munich. Later, and up through the final phase of Klein's highly successful career in Göttingen, he continued to cultivate close relations with his longtime allies in Erlangen.[5]

3.2 Academic Antisemitism

These events from the early 1870s led to an unusual situation in Erlangen. During an era when very few German Jews could hope to attain a professorship in Germany, both mathematicians on the small faculty at Erlangen University were of Jewish background. This unusual circumstance certainly did not go unnoticed at the time, and the present section attempts to gauge the effects of academic antisemitism on their careers. Unlike Max Noether, who remained a non-practicing Jew all his life, Paul Gordan converted to Christianity at the age of 18.[6] Still, in the eyes of many, a baptized Jew was not to be confused with a "real German."

This pervasive attitude surely helps to explain why both Gordan and Noether never had a chance to leave Erlangen. In Gordan's case, he may have been quite content to stay in Erlangen since he was already a full professor, but Noether, as an associate professor, could hardly feel the same way. Yet he was passed

---

[5]Over the course of their friendship, Klein and Noether exchanged some 340 letters, from which 280 are still extant in Klein's estate (SUB Göttingen).

[6]A finding due to Cordula Tollmien, who kindly sent me a copy of Gordan's baptismal certificate dated July 1857. Tollmien points out that Gordan's baptism took place before he began his academic studies, though nothing is known about his motives at this time.

Figure 3.2: Max Noether (Auguste Dick Papers, 13-1, Austrian Academy of Sciences, Vienna)

over on numerous occasions; in some cases Klein informed him in advance that certain localities were simply opposed to any and all Jewish candidates. Over time, Noether came to realize that his best chance for promotion would likely come if Gordan were to receive an outside offer; that was Klein's frank opinion, too. When a mathematics professorship opened in Tübingen in 1885, Noether hoped this might indeed transpire.[7] A short time before, Max Noether's friend and collaborator, Alexander Brill, was appointed to a newly established second chair there, a situation that lifted Noether's hopes that Gordan might well be chosen. Instead, however, Gordan's candidacy received no serious consideration at all, as Brill informed him in a letter from July 1, 1885:

> You should know that it was not the faculty or the Senate that blocked Gordan's candidacy, nor was it the chancellor nor the government: the entire country [meaning the state of Württemberg] is currently of such

---

[7]This was the position formerly occupied by Paul Du Bois-Reymond, who one year earlier accepted an offer from the Technische Hochschule in Berlin. Hermann von Stahl from Aachen Institute of Technology was eventually appointed his successor in Tübingen.

a mind that a professor of Jewish origin in Tübingen is impossible. This can and will change, but as a newcomer I am unable to make the first breach in this prejudice. [Seidl, et al. 2018, 23]

Brill gave no clear indications as to what was behind this disaffection for Jewish candidates. He merely stated that this specific appointment had caused a great deal of controversy because of differences between Paul Du Bois-Reymond, the previous chair holder, and the faculty, a circumstance that obviously had no bearing on the issue of antisemitism. More than likely, Brill alluded to this merely in order to explain why he, a newcomer, had only limited influence on the faculty's decision. As a matter of fact, before this time only one mathematician of Jewish origin, Sigmund Gundelfinger, had ever been a member of the Tübingen faculty.[8] It should be noted that Brill's prediction, according to which future prospects for Jewish mathematicians would improve in Tübingen, never materialized. Although his friendship with Max Noether apparently remained firm over the years, his general view of German Jewry became increasingly hostile, reflecting opinions held by conventional antisemites. On January 5, 1914, not long before the outbreak of World War I, Brill wrote this entry in his diary:

The effect of the Jews on Germanic peoples is like alcohol on the individual! In small doses they are stimulating and invigorating, but in large quantities devastating like poison. The organism of our people requires time to assimilate them. Therefore they should be warded off because otherwise the flood from the east threatens to destroy the body of the people, like aphids attacking a plant, which will then perish. Fend them off! They know nonetheless how to smuggle themselves in. [Seidl, et al. 2018, 23-24]

This theme of Germania as the victim of merciless and conspiring Jews, who threatened to invade the young nation from the East, would become a standard trope in the period after the monarchy fell in November 1918. The fact that Brill had already adopted this viewpoint even before the outbreak of the Great War suggests how deep-rooted these types of fears must have been among Germany's educated classes.

Conditions in Erlangen during the Wilhelmian age may have been more liberal, at least in some academic circles, but German Jews who managed to attain professorships were acutely aware that their presence on university faculties was rarely welcomed [Kaplan 1991, 137–150]. In some disciplines, classical philology being a noteworthy example, scholars of Jewish origin had virtually no chance of advancement. Mathematics, on the other hand, was long seen as a field in which

---

[8]Coincidentally, Gundelfinger had studied under Clebsch and Gordan in Giessen and, like them, he worked mainly on invariant theory and its application to algebraic curves. After taking his doctorate in Giessen in 1867, he habilitated two years later in Tübingen, where he was appointed associate professor in 1873. Six years later he joined the faculty at the Technical University in Darmstadt as a full professor.

high-quality research was recognized objectively and judged accordingly. If that was the ideal, then the reality was very different indeed.

In today's universities, mathematics is strongly allied with the natural sciences, in part due to the current importance of applied mathematics. Historically, however, these relationships were by no means self-evident. During the nineteenth century, the ties between mathematics and the human sciences were, in some ways, the stronger ones. First, it should be remembered that the research interests of most mathematicians at the German universities were devoted to some branch of pure mathematics. It was not until the advent of the twentieth century that applied fields began to receive strong attention. Second, throughout most of the nineteenth century, humanists and natural scientists were colleagues in a single philosophical faculty. Mathematicians could therefore interact just as easily with philologists and philosophers as with their colleagues in astronomy and physics. Third, and perhaps most important, it was mainly the humanists who set the tone at faculty meetings and in broader forums outside the university proper. The most prominent among them spoke as *Kulturträger*, an elite class of intellectuals often called "Mandarins" (*Bonzen*). This group reached its zenith during the last decades of the Wilhelminian era. Its demise began with the fall of the German Reich, accelerating as Germany descended into Nazism; this familiar story is described and documented in detail in [Ringer 1969]. Looking backward to the early decades of the nineteenth century, mathematicians often had a stronger affinity for idealistic philosophy than for the materialism many identified with the natural sciences. The latter fields had, in any case, a lower status than the human sciences, and since mathematicians saw themselves as purveyors of pure knowledge they naturally followed the lead of their colleagues in classical philology, who were the first to establish research-oriented seminars.

One seminar that was particularly influential for physics and mathematics was founded in 1834 in Königsberg [Olesko 1991]. Initially, this seminar was under the direction of the physicist Franz Neumann and the Jewish mathematician Carl Gustav Jacob ("Jacques") Jacobi, who later went to Berlin in 1844. Jacobi, the son of a Potsdam banker, became a model figure for numerous Jewish mathematicians who pursued careers in Germany after him, one of whom was Leo Koenigsberger, Jacobi's biographer [Koenigsberger 1904]. Koenigsberger delivered a speech in honor of his hero at the Third International Congress of Mathematicians, which was held in Heidelberg in 1904, one hundred years after Jacobi's birth. He ended this oration by proclaiming: "We are all students of Jacobi." This, of course, was an exaggeration, and a well-known German mathematician[9] wrote to Felix Klein, wondering how anyone could make such a claim, overlooking Gauss and Riemann [Rowe 2018b, 28-29].

Koenigsberger, who was himself a baptized Jew, knew very well that Jacobi had to undergo baptism in order to pursue an academic career. Yet, this circumstance went unmentioned in his biography, which even ignored Jacobi's Jewish

---

[9]He was Klein's former student Walther von Dyck; see [Hashagen 2003]

origins.[10] This reticence to address the "Jewish question" stands in stark contrast to Koenigsberger's autobiography [Koenigsberger 1919], which he published in 1919. There he went into detail about the painful conflict that many young Jews had to face in deciding whether they should be baptized. This was less the case for Koenigsberger himself, but the question plagued his longtime friend Lazarus Fuchs, with whom he studied in Berlin in the 1860s:

> Fuchs and I had to ask ourselves whether we should sacrifice our entire scientific life and existence because of the prevailing, narrow-minded views of the government or, after we had stripped away all religious prejudices, instead convert to Christianity. Fuchs had already let three years go by in hesitation and indecision because he had to take diverse concerns of his family into consideration, whereas I was free from such bonds, coming from a household that was hardly religious, and so my firm urging on Fuchs, who had been fearful and timid throughout his life when making important decisions, resulted in his future, too, being saved. [Koenigsberger 1919, 18]

If one compares this passage with Koenigsberger's Jacobi biography, in which he completely avoided the topic of the Jewish question, one can hardly escape the impression that this topic was completely taboo during the Wilhelmian era. It seems likely that the critics of this state-sanctioned form of academic antisemitism only felt free to speak about it in private circles. In the context of careers in mathematics, one is reminded of the long struggle faced by Gauss' student, Moritz Abraham Stern [Schmitz 2006]. Stern refused to be baptized and consequently had to wait 30 years before he was appointed full professor in Göttingen in 1859. Stern's earlier appointment to an associate professorship in September 1848 came at a time of symbolic significance for German Jews. Many had pinned their hopes for social progress and true emancipation on the success of liberalism, a movement that lost momentum after 1848 and then fell into decline.

Yet with increasing trends toward secularization, compulsory baptism gradually declined as well. Nevertheless, subtler forms of academic antisemitism continued to prevail at German universities. One only rarely finds references to the religion or ethnic background of the candidates in official documents concerning appointment procedures, but these aspects surfaced quite often in private correspondence. Such considerations often came into play, but one should not overlook another factor that obviously discouraged many talented young men who might have dreamed of a life as a university professor, namely the competition for such positions. In mathematics, there were roughly only a hundred full professorships in all the German states combined! Not until the latter half of the twentieth century did the number of positions increase dramatically. Little wonder, then, that

---

[10]The only place in which Judaism comes up at all is in a letter from F.W. Bessel to C.F. Gauss, written shortly after Jacobi came to Königsberg as a private lecturer. Bessel described him as "very talented" but also tactless. He also believed he had heard that "his father was a Jew and money changer in Potsdam" [Koenigsberger 1904, 27].

many Jews chose to study law or medicine, fields they could practice privately, rather than trying to pursue a university career [Richarz 2015]. By the second half of the nineteenth century, the natural and engineering sciences were also opening new doors, whereas the humanities continued to prepare most of the candidates for the teaching profession and other civil service positions.

University careers in mathematics thus posed daunting challenges, though this field offered better prospects for Jews than did many others. Their chances, in fact, depended heavily on attitudes regarding the "Jewish question" within the respective faculties. The fact that Göttingen later attracted many young Jewish mathematicians, especially after 1900, had much to do with the comparatively liberal atmosphere that prevailed there ([Rowe 2004], [Rowe 2018a, 171 –232]). Young Jews, like Richard Courant and Max Born, were encouraged by witnessing the extremely close friendship between David Hilbert and his Jewish colleague Hermann Minkowski, who had studied together in Königsberg. By the end of the Weimar period, Göttingen was seen in the eyes of the antisemites as the stronghold for a Jewish conspiracy within German mathematics and physics. Albert Einstein's numerous enemies also counted him as part of the "Jewish network" centered in Göttingen.[11]

As noted already, the careers of Gordan and Noether were closely intertwined, despite notable differences in their temperament and outlook. Both deferred to Felix Klein as the acknowledged new leader of the Clebsch school. One of its younger members, Ferdinand Lindemann, studied briefly with Clebsch before completing his education under Klein in Erlangen. Soon thereafter, he made a name for himself by publishing the first volume of Clebsch's lectures on geometry [Clebsch/Lindemann 1876]. During the period from 1880 to 1883, Lindemann was professor of mathematics in Freiburg, where he solved one of the oldest and most famous problems in mathematics by proving the impossibility of squaring the circle. This had long been suspected, but Lindemann found a way to prove that π is a transcendental number and thus can never arise as the solution of a conventional algebraic equation. Not long afterward, he succeeded Heinrich Weber as professor of mathematics in Königsberg, a major breakthrough for him as well as for members of the Clebsch school. Up until this time, none had managed to gain a professorship at a Prussian university, as these chairs invariably went to graduates from Berlin University or their allies.

Lindemann's departure from Freiburg momentarily raised Max Noether's hopes that he might be chosen to succeed him. Klein, who was eager to help, made inquiries, and then sent Noether his personal assessment of his friend's chances both in Freiburg and elsewhere:

> I don't yet have any definitive news from Freiburg, but I'm afraid you have little to hope for there, despite the fact that I once again vigorously recommended you as did [Theodor] Reye.[12] And Tübingen, where a new

---

[11]See [Rowe 1986] and [Rowe 2018a, 258–261].
[12]The chair went to Jacob Lüroth, a close friend of Noether.

full professorship has in fact been approved, will also bring you nothing, because I heard from there in no uncertain terms that your confession would constitute an obstacle.[13] These are two pieces of bad news that I take no pleasure in writing. It would seem that Erlangen remains the place that offers you the best prospects. If only there were some desire to bring Gordan somewhere else! But in the eyes of the world he looks like a permanent Erlangen fixture. That's what happens when one remains isolated and stops seeking new working relationships. I lectured him about this over the Easter holidays, but he has little desire to change.[14]

A similar situation arose four years later in Giessen, where Moritz Pasch (1843–1930) was a member of the search committee called upon to fill a vacancy there. Pasch, himself of Jewish origin, informed Klein about final candidates for this position, none of whom were of Jewish extraction.[15] He explicitly pointed this out by noting that he would have otherwise recommended Max Noether and Klein's student Adolf Hurwitz.

When Klein learned in 1888 that Noether would be promoted to full professor in Erlangen, he wrote him to express his congratulations, but also his personal relief that this ordeal was finally about to end.[16] Three months later, he read about Noether's new appointment in the newspaper.[17] Paul Gordan's last best chance to attain a more prominent position came five years after this. In 1893, Klein declined a call from the Ludwig Maximilians Universität in Munich, at which time he was asked to recommend candidates for this professorship [Tobies 2019, 337–338]. Instead of Gordan, Klein named Lindemann, who would spend the remainder of his career in Munich. Gordan was most unhappy about this turn of events, which temporarily strained his relationship with Klein, who was mainly focused on trying to bring David Hilbert to Göttingen.[18] His plans for doing so came to fruition one year later.

And so it transpired that the University of Erlangen, with its two mathematicians of Jewish origin, remained a singularity in the mathematical landscape of Germany. By the early 1890s, Max Noether turned his attention more and more to the broader mathematical literature, whereas his colleague Paul Gordan continued to pursue his special research interests. Aside from Emmy Noether, Gordan supervised only one other doctoral student, the American Harry W. Tyler. How that came to pass reflects once again the close connections between Klein and his two friends in Erlangen.

---

[13]This concerns the position that ultimately went to Alexander Brill, who remained in Tübingen until his retirement in 1918.

[14]Klein to Max Noether, May 29, 1883, Klein Nachlass, SUB Göttingen; reproduced in [Bergmann/Epple/Ungar 2012, 189].

[15]Pasch an Klein, 1887, Nachlass Klein, SUB Göttingen.

[16]Klein to Max Noether, January 31 1888, Nachlass Klein, SUB Göttingen.

[17]Klein to Max Noether, April 24, 1888, Nachlass Klein, SUB Göttingen.

[18]The tensions between Klein and his longtime Erlangen friend are apparent in Gordan to Klein, December 5, 1892, Nachlass Klein, SUB Göttingen.

After studying at MIT, Tyler arrived in Göttingen in 1887 during a period when several young American mathematicians were flocking to Klein's courses [Parshall/Rowe 1994]. During his first year of study, Tyler thought about moving on to Berlin, but decided instead on Erlangen. He reached this decision following a conversation with Gordan, who had come to visit Klein in Göttingen. Presumably these two friends and former collaborators encouraged Tyler to test the waters in Erlangen, and so the latter took the plunge. One of the Americans who came to Göttingen at the same time as Tyler was William Fogg Osgood from Harvard, later to become a fixture of its mathematics faculty. Osgood remained in Göttingen after Tyler left, but he was also curious to learn about mathematical life in Erlangen, which led to an interesting correspondence with his American friend. Tyler not only offered advice about how Osgood should prepare in the event he might decide to write his dissertation in Erlangen, he also contrasted his new academic environment with the one he experienced one year earlier. "I'm very glad I went to Göttingen first," he wrote, "a first semester here would have been a mournful experience for me. ...I understand in a measure the superiority which Klein and Gordan each credited to himself or to the other when I met the latter in Göttingen, and although I have at present more admiration for Kl[ein] I feel better satisfied not to have worked with him alone" [Parshall/Rowe 1994, 229].

Tyler had not come to Erlangen empty-handed, as Klein had given him a thesis topic closely connected with the coursework Tyler did with him. This concerned Abelian integrals over a ground curve whose only singularities were double points. Meanwhile, however, Tyler took up an independent study on resultants under Gordan, never imagining that this topic might be a suitable dissertation topic. Yet little more than one month after his arrival in Erlangen, Gordan informed him that he could use it for his doctoral thesis. This came as a big surprise to Tyler, who knew very well that he was only grazing in pastures that were long familiar to experts in algebra. He reported to Osgood that the coursework in Erlangen was in no way comparable to the offerings in Göttingen, but as compensation he had nearly daily contact with Gordan and Noether. Since the latter was unable to walk to classes due to his physical handicap, Harry Tyler and another student took turns transporting him from his home to the university in his wheel chair.

A few months later, after Osgood inquired again, Tyler offered a quite lengthy estimation of the pros and cons of study in Erlangen as opposed to Göttingen. In his view, Americans who had three years to pursue graduate studies in Germany would be well advised not to spend the entire time studying under Klein, who though a brilliant lecturer was less than ideal as a role model. Klein offered his auditors the chance to gain a sweeping overview of major fields of research, but he only occasionally discussed a complex theorem in any detail. As Tyler saw it, "this broad view of things is very attractive and something that any student may go to Göttingen for, but there seems to me danger that in attempting to follow in Klein's direction [one] will produce only rubbish." He also underscored another drawback for those who chose Klein as their thesis advisor, namely, "[so] busy a man cannot and will not give a student a very large share of his time and attention; so too

he will not study out of interest himself especially in the painstaking elaboration
of details, preferring to scatter all sorts of seed continually and let other people
follow after to do the hoeing" [Parshall/Rowe 1994, 231–232].

In this respect, Tyler considered the conditions in Erlangen far better suited
for dissertation work. He had now completed his own thesis and assured Osgood
that he would easily be able to do the same during his final year of study. "Anyone
coming here from Klein," he wrote, "would be sure to look at mathematical things
from a new standpoint and ... would be practically certain of a degree of interest
and attention almost out of the question in Göttingen, and especially valuable
when one is beginning original work. I have been and am still embarrassed by the
opportunities." This did not mean, however, that Tyler always found it easy to
stay on good terms with his two teachers, who were temperamentally extremely
different. "Both men," he pointed out to Osgood, "are so peculiar and so irrecon-
ciliable that ... [personal relations] must be cultivated with some tact especially if
one tries to divide his attentions equally. So far as I know N[oether] like G[ordan]
confines himself to pure mathematics – though he studied physics with Kirchhoff
– and both I think run to *Tiefe* [depth] rather than *Breite* [breadth], as compared
with Klein. If they have that much in common, that's about all. G[ordan] is
outspoken, irascible, exasperating, violent; N[oether] is taciturn, serious equable,
patient" [Parshall/Rowe 1994, 232].

Gordan and Noether were, indeed, very different types of personalities; though
they apparently never became close friends, their relationship seems to have remained fairly collegial. A contrary opinion was once expressed by Erhard Schmidt,
who taught in Erlangen for just one year, arriving in 1910 as Gordan's successor.
During a celebration in his honor in Berlin, Schmidt recalled that time from four
decades earlier. He imagined this would be a rich and rewarding experience during
which he would enjoy a good deal of contact with Gordan and Noether. After he
arrived, however, he noticed:

> ... a small difficulty in that these two luminaries had not been able to
> stand one another for thirty years – and with a deep antipathy that
> typically develops in smaller cities, where one tends to encounter the
> object of one's anger on nearly every street corner. [Schmidt 1951, 22][19]

Schmidt was a witty man, and on such an occasion he no doubt enjoyed embel-
lishing on such anecdotes from his long career.

Tyler ended his long letter to William Osgood by offering the following summary advice:

> ... come here if you want ... detailed work in pure mathematics. If you
> want to work especially with Gordan I wouldn't suggest any prepara-
> tion unless the first volume of his book [Gordan 1885/1887]. If you had
> anything underway very likely it wouldn't interest him. For Noether,
> on the other hand, I think it would be worthwhile to have something

---

[19]Thanks to Reinhard Siegmund-Schultze for pointing out this source.

yourself to propose – in Abelian functions if you like or any of his sub-jects that you know from the Annalen as well as I could tell you. I wouldn't advise you to come unless you feel sure your tastes lie in these directions. [Parshall/Rowe 1994, 232 233]

Osgood took his friend's words to heart. The following semester, he arrived in Erlangen already prepared to write his dissertation on a topic in Abelian functions, which he completed the following academic year. His chosen mentor, not surprisingly, was Max Noether. Presumably Noether would have also supervised Tyler's dissertation on the topic Klein had given him, had not his impulsive col-league jumped into the fray with his own proposal for the American. Not that Paul Gordan was keen to supervise students' doctoral dissertations, a chore he usually left to Noether. The latter served as mentor to eighteen doctoral stu-dents over the course of his career, compared with two for Gordan. Nevertheless, Gordan's voluble personality and eccentric mannerisms made him highly popular with students as well as colleagues. Osgood, too, learned a great deal from Gordan during the year he spent in Erlangen. When he returned to Harvard, he published [Osgood 1892], an expository paper on the German symbolic methods for cal-culating invariants in an effort to make these accessible to the English-speaking mathematical world.

Tyler's letters to Osgood were written a decade before Emmy Noether began auditing courses at Erlangen University; still, there is every reason to believe that mathematical life there had changed very little in the meantime. No doubt she and her brothers took turns wheeling their father from home to campus on a daily basis. Surely he guided her studies during these early years, and at some point along the way she began reading [Brill/Noether 1894], the monumental historical study on algebraic functions that he and Alexander Brill wrote for the newly founded German Mathematical Society. In the following section we sketch some of the most important early influences on Emmy Noether's life and how these shaped her outlook as a mathematician.

3.3 Emmy Noether's Uphill Climb

Mathematical talent, if it is to be realized, requires ample amounts of nature *and* nurture. In the case of the Noether family, this leads to an obvious question. What kind of influence did Max Noether exert on his two mathematically gifted children, Emmy and Fritz? Although barely any documentary evidence survives that provides information about the encouragement and/or pressure they might have received from their father, it is striking that both of his children went their own way in mathematics; and both chose very different paths. It may well be that Max was an ambitious and demanding father, fulfilling what was then the standard role in those days. Still, one finds nothing that suggests his parenting had an adverse effect on Emmy or Fritz.

Figure 3.3: Emmy Noether and her three brothers, l. to r., Alfred, Fritz, and Gustav Robert (Auguste Dick Papers, 12-14, Austrian Academy of Sciences, Vienna)

Emmy graduated from the local high school for girls in 1897, after which she began preparing to take the Bavarian state examinations that would qualify her to teach French and English at similar schools for young women. Two years later, she passed those exams with flying colors, receiving the highest grade (1,0) in all parts except for "classroom teaching," for which she only received a grade of 2,0. As Auguste Dick commented, the lower grade was no doubt justified, as "even later, as a university lecturer, she would not have done better" [Dick 1970/1981, 1981: 12]. Having attained this higher qualification, Noether was now allowed to audit courses at the university. At the same time, she and her brother Fritz also studied together in preparation for taking the *Abitur* examination. After both passed this difficult test in the summer of 1903, they could make plans to enter Erlangen University. As described in the previous chapter, for Emmy this came as a true stroke of good luck, as it was in that very year that the state of Bavaria granted women the right to matriculate at institutions of higher education. She opted, nevertheless, to first study mathematics at the Prussian university in Göttingen, possibly following her father's advice.

As a close friend of Felix Klein, head of Göttingen's stellar mathematics faculty, Max Noether knew that there was no better place on earth for an aspiring young mathematician to take wing. In any event, she enrolled at the Georgia Augusta in the winter of 1903, but only as an auditor since women were still not allowed to attend Prussian universities as regularly matriculated students. This bold venture did not go well, however, possibly due to Emmy's excessive ambition. She not only took courses offered by Klein, Hilbert, and Minkowski, but also attended the lectures of Otto Blumenthal, and the astronomer Karl Schwarzschild. It was all too much for a first semester student. When she returned home at the end of these studies, she was seriously ill. Her family then sent her to a quiet place in the countryside where she could rest and recover. This period of convalescence proved to be just what Emmy Noether needed, and so she afterward took up her studies again, but this time in Erlangen, where she could live at home.[20]

Emmy, Fritz, and their brother Alfred all attended the local university at this time; Alfred completed his doctorate in chemistry in 1909 (Fig. 3.3). From 1904 onward, Emmy and Fritz Noether spent five semesters together at the University of Erlangen before Fritz decided to continue his studies in Munich. Both were friends of Hans Falckenberg, whose father, Richard Falckenberg, was professor of philosophy in Erlangen [Dick 1970/1981, 1981: 16]. Hans and Fritz served together during the war, called up from their respective teaching posts in Brunswick and Karlsruhe. Fritz Noether's academic career was more conventional than his friend's, as the young Falckenberg began, like Karl Weierstrass, by studying law, before switching to mathematics. Although he was no Weierstrass, he found the support he needed in Erlangen, not least from Emmy Noether.

Hans Falckenberg's doctoral dissertation dealt with a topic in analysis closely related to the research interests of Erhard Schmidt, who succeeded Gordan in 1910, but then left for Breslau after just one year. Ernst Fischer, who was well versed in analysis, thus served as Falckenberg's official advisor. At first glance, one would never imagine that Emmy Noether would have had a role in this: what did she know about the "branching of solutions of nonlinear differential equations"? Yet reading her curriculum vitae at the time she finally habilitated in Göttingen in June 1919, we find at the very end: "Finally, I would like to mention that, beyond those mentioned above, another Erlangen dissertation by H. Falckenberg was prompted by me .... This concerns the investigation of reality relations in connection with work by Schmidt on integral equations" [Koreuber 2015, 21]. Clearly, Emmy Noether's eyes were wide open after she completed her own doctorate, and in this case Hans Falckenberg was the beneficiary of her newfound expertise, for which he thanked her profusely. After his stint as a private lecturer, in 1922 Fal-

---

[20]Max Noether to David Hilbert, April 27, 1904, and Max Noether to Felix Klein, November 14, 1904, cited in [Tollmien 1990, 160].

ckenberg was appointed associate professor in Giessen; he became a full professor there in 1931.

Both Emmy and Fritz also surely knew Emil Hilb, who grew up in Württemberg as the son of a Jewish merchant. His family later moved to Bavaria, where Hilb attended the Realgymnasium in Augsburg. Afterward he studied in Berlin and Göttingen, before taking his doctorate in 1903 in Munich. Hilb then got a teaching position at his old secondary school in Augsburg, where he worked for three years until Max Noether discovered his mathematical talent. Noether brought him to Erlangen in 1906 as his assistant. Two years later, Hilb became a private lecturer at the University of Erlangen, before being appointed to an associate professorship in Würzburg in 1909.

As a mathematician, Emmy was much closer to her father than was her brother, who turned to applied mathematics. Fritz Noether's broad interest in technical applications was already noticed by his professor of physics, Arnold Sommerfeld, who gave him a free hand to complete the last part of the four-volume study [Klein/Sommerfeld 1910] on the theory of gyroscopes. Emmy, too, studied physics in Erlangen, however her early interests centered on pure mathematics, especially problems in classical invariant theory, which she learned while working under the supervision of Paul Gordan. Her *Doktorvater* was an impulsive man with a sarcastic sense of humor. Hermann Weyl thought of him as a typical representative of the 1848 generation, the eternal "Bursche" who went around in a nightshirt that smelled of beer and tobacco ([Weyl 1935]). Gordan's mathematics somehow matched his moody personality, too. He was one of the last great algorists; his contemporaries knew him as the "King of Invariants," the man who proved the finite basis theorem for binary algebraic forms by showing, in principle, how to calculate a basis for the invariants and covariants of any system of forms of arbitrary degree. His colleague, Max Noether, was familiar enough with invariant theory, but from a broader standpoint, not as a specialist like Gordan.

Max Noether presumably held strong views when it came to the mathematical education of his children. The fact that neither Emmy nor Fritz pursued algebraic geometry, their father's main field of expertise, would seem telling. This could very well reflect a conscious decision on the part of Max Noether to let both of them strike out on their own. In 1909, Fritz completed his doctorate in Munich under Aurel Voss with a thesis in kinematics (rolling motions of a sphere on surfaces of revolution). He spent the next year assisting Arnold Sommerfeld in finally completing the final volume of [Klein/Sommerfeld 1910]. This paved the way for a similar one-year appointment in Göttingen, where Noether assisted Carl Runge, who also took him in as a boarder. During this year, Fritz Noether befriended Hermann Weyl, a talented analyst and the rising young star in Hilbert's fast-growing school. Weyl probably only met Fritz's sister much later, when he was a visiting professor in Göttingen during the winter semester of 1926/27. Following this two-year stint as a post-doc, Fritz Noether gained a position as assistant to Karl Heun at the Karlsruhe Institute of Technology (KIT). Soon thereafter, he submitted his habilitation thesis, and in the summer of 1911 he became a private lecturer at

KIT. Noether served on the Western front until he was wounded there, after which he was employed to do research on ballistics. Once the war ended, Fritz Noether returned to KIT, where he was promoted to an associate professorship. After three years, he took a leave of absence to work at the Siemens-Schuckert electrical company in Berlin, after which he obtained a full professorship in 1922 at the Breslau Institute of Technology. He there joined Emmy Noether's protégé Werner Schmeidler, who had succeeded Max Dehn just one year before (see Section 2.2).

On 20 December 1911, only months after his appointment as a lecturer in Karlsruhe, Fritz Noether married Regina Maria Würth, the daughter of a customs official. She grew up in a large Catholic family, originally from the Black Forest region in southwestern Germany, and her husband converted to Catholicism around the time they wed. Not long thereafter, their two sons were born: Hermann and Gottfried Noether. Both went on to successful careers in the United States: Hermann (who changed the spelling of his name to Herman) as a chemist and Gottfried as a statistician. Fritz Noether was still only a private lecturer at the time he married, so he presumably relied on subsidies either from his mother or possibly from wealthy members on her side of the family. Emmy later received financial support from two wealthy uncles, Wilhelm and Paul Kaufmann, who lived in Berlin [Dick 1970/1981, 8].

Since Emmy had every opportunity to learn algebraic geometry from her father, he no doubt wanted her to explore other terrain, and in Erlangen this left only one alternative: she would learn to do invariant theory under Paul Gordan. Indeed, her doctoral dissertation was written entirely in his algorithmic style. In her younger years, she clearly had a gift for calculation, as the task Gordan gave her was a truly daunting one: to develop the complete invariant theory for ternary forms of degree four. Whether Gordan had ever attempted this is hard to say, but one can safely say that Emmy Noether's work was the last word with regard to this problem. She was able to construct 331 associated covariants, though who will want to say that her list is really complete? After finishing this ambitious work and publishing it in *Crelle* [Noether 1908], she passed her oral exam "summa cum laude." If she herself later dismissed her dissertation as a piece of juvenilia (she once called it "dung"), it should not be overlooked that this study demonstrated her ability to manipulate algebraic expressions of a most unwieldy variety. Indeed, in her youth the "mother of modern algebra" was a first-rate algorist, and though her later publications were models for abstract clarity, these works were always grounded in a thorough mastery of relevant special cases. Noether's mature studies reflected her deep interest in substantive mathematics; she was never content with empty abstraction.

During the following eight years, Emmy Noether continued to mature as a mathematician, particularly under the influence of Gordan's successor, Ernst Fischer, who arrived in Erlangen in 1911. Fischer grew up in a musical family; his father was a composer and taught at the Vienna Academy and his maternal grandfather was also a musician. Ernst Fischer studied mathematics at the University of Vienna under Franz Mertens, who was a leading expert in invariant theory. Before he came to Erlangen, Fischer taught in Brünn. Few sources have survived that shed real light on Emmy's intellectual development during the period 1908 to 1914, but these were clearly important years during which her mathematical capabilities became ever more apparent to those who knew her best. Although she undoubtedly spent a good deal of time at home studying the books in her father's library, she also continued to take part in mathematical life at the university. The directors of the mathematics seminar in Erlangen – Max Noether as well as Gordan's successors, Erhard Schmidt and Ernst Fischer – called on her regularly to assist them with seminar lectures and practice sessions. Gradually, she began to make a name for herself as an expert in *modern* invariant theory, not the old-fashioned version she had learned under Gordan and by reading [Gordan 1885/1887]. This new direction in her research actually dated back to Hilbert's papers from the late 1880s, which reoriented research in invariant theory.

It was through her interactions with Fischer that Noether discovered her real strengths and interests in abstract algebra. Noether's first biographer, Auguste Dick, aptly described Ernst Fischer's impact on her subsequent career in these words:

> With him she could "talk mathematics" to her heart's desire. Although both lived in Erlangen and saw each other frequently at the University, a large number of postcards exist from E. Noether to E. Fischer, containing mathematical arguments. Looking over this correspondence, one gets the impression that immediately after a conversation with Fischer, Emmy Noether sat down and continued the ideas discussed in writing, whether so as not to forget them, or whether to stimulate another discussion. Ernst Fischer has succeeded in carefully preserving these communications through all the havoc of war. The correspondence extends from 1911 to 1929 and is most frequent in 1915, just before Emmy Noether moved to Göttingen and Ernst Fischer was drafted by the military. There can be no doubt that it was under Fischer's influence that Emmy Noether made the definite change from the purely computational distinctly algorithmical approach represented by Gordan to the mode of thinking characteristic of Hilbert. [Dick 1970/1981, 23]

Unfortunately, almost nothing seems to have survived from this correspondence between Fischer and Noether, as these documents would surely have provided many insights into their mathematical discussions and growing relationship during the critical transitional years before Emmy left Erlangen for Göttingen. Nevertheless, her publications from this period contain numerous hints and references to ideas she discussed with Fischer, so that by scrutinizing these carefully it becomes possible to draw a general picture of their mutual interests. Looking backward, Emmy always emphasized that Fischer's presence in Erlangen had opened her eyes to a very different world of mathematical ideas, a realm that had been closed to her as a student of Gordan and her father.

3.4 Classical vs. Modern Invariant Theory

Paul Gordan's approach to invariant theory still dominated the field up until the early 1890s, when David Hilbert began recasting the entire subject in terms of general theorems for algebraic systems. Between 1888 and 1890, Hilbert succeeded in generalizing Gordan's finiteness theorem from binary forms to families of forms in any number of variables. Initially, Gordan followed Hilbert's work with enthusiasm, but by 1890 he was going around telling anyone who would listen that Hilbert's approach to invariant theory was "theology not mathematics" [McLarty 2012]. Max Noether certainly heard this pronouncement from Gordan himself, since he recorded it in [M. Noether 1914]. Part of what Gordan disliked stemmed from the fact that Hilbert relied heavily on existence arguments. Rather than exhibiting an algorithm or general procedure for constructing a finite basis, he showed that such a basis had to exist out of sheer logical necessity.

These new ideas departed radically from the traditional ground rules for research in invariant theory adopted by Emmy Noether's mentor, Paul Gordan, and other practitioners. Little wonder that the older man felt offended and scoffed at Hilbert's work, labeling it a pretentious new "theology." Felix Klein took a decidedly more open-minded attitude. When he received the manuscript for [Hilbert 1890], Klein wrote back one day later: "I do not doubt that this is the most important work on general algebra that the Annalen has ever published."[21] At the time, this conflict had no major ramifications, yet it clearly foreshadowed a highly significant methodological shift that played a central role in Emmy Noether's education as a mathematician. For her mentor, Paul Gordan, mathematics was essentially a playground for manipulating complicated algebraic formulas. As Gordan described himself in a letter to Klein: "I can only learn something that is as clear to me as the rules of the multiplication table."[22]

Two years later, however, Hilbert managed to find a new proof that was constructive in spirit. Although this route to his finiteness theorems was more complicated, it carried a major new payoff, namely "the determination of an upper bound for the degree and weights of the invariants of a basis system."[23] When Minkowski learned about Hilbert's latest triumph, he fired off a witty letter congratulating his friend back in Königsberg:

> I had long ago thought that it could only be a matter of time before you finished off the old invariant theory to the point where there would hardly be an i left to dot. But it really gives me joy that it all went so quickly and that everything was so surprisingly simple, and I congratulate you on your success. Now that you've even discovered smokeless gunpowder with your last theorem, after Theorem I caused only Gordan's eyes to sting anymore, it really is a good time to decimate the

---

[21]Klein to Hilbert, 18 Feb. 1890, [Frei 1985, 62].
[22]Gordan to Klein, 24 February 1890, [Frei 1985, 65].
[23]Hilbert to Klein, 5 January 1892, [Frei 1985, 77].

fortresses of the robber-knights [i.e., specialists in invariant theory] –
[Georg Emil] Stroh, Gordan, [Kyparisos] Stephanos, and whoever they
all are – who held up the individual traveling invariants and locked them
in their dungeons, as there is a danger that new life will never sprout
from these ruins again.[24]

It would be a mistake to imagine that the clash between Gordan and Hilbert was
merely a momentary episode that pitted two headstrong personalities against one
another. For Hilbert, much was at stake methodologically. In 1890, Gordan was
in no position to contest Hilbert's formalist views, but thirty years later the Dutch
intuitionist L.E.J. Brouwer launched a serious attack on them.

Hilbert later summarized his new results in the classic paper [Hilbert 1893],
his final contribution to invariant theory. Here he adopted an even more general
standpoint by treating invariant theory as a special case of the general theory of
algebraic function fields, while underscoring the close analogy between the latter
and algebraic number fields. In his introduction, he set down five fundamental
principles which could serve as the foundations of invariant theory. The first four
concerned "elementary propositions of invariant theory," whereas the fifth principle
asserted the existence of a finite basis (or in Hilbert's terminology a "full invariant
system"). This highly abstract formulation would later become a hallmark for
almost all of Hilbert's work under the rubric of the modern axiomatic method.
Some two decades later, Emmy Noether would take up this thread of ideas again,
along with other strands of algebraic work developed by Richard Dedekind and
Ernst Steinitz. Inspired by her many conversations with Ernst Fischer, she was
already steeped in this world of mathematical thought when she left Erlangen for
Göttingen in the spring of 1915.

A brief, but telling glimpse of Noether's research interests in the period before her departure can be seen from the text she prepared from a talk delivered
at the September 1913 meeting of the German Mathematical Society in Vienna
[Noether 1913]. She began with two remarks, the first being that the questions presented originally arose through discussions with Fischer. Secondly, she noted that
certain special cases in a single variable had already been answered by Steinitz,
but that his methods were also valid for more general fields. Emmy Noether would
later take full advantage of these ideas and results, but for her Vienna lecture she
dealt with fields of rational functions defined over the more restricted terrain of
number fields. Her preliminary results and long-term goals concerned finiteness
theorems for these rational function fields. This involved finding conditions that
lead to a finite set of basis functions for determining all others. These types
of finiteness issues – all quite closely connected with Hilbert's style of algebraic
research – would occupy Noether for the next several years.

In 1914 Noether corresponded with Hilbert about this and other work, including [Noether 1915], which elaborated on results she announced in her Vienna lecture. In May, she sent him the manuscript for this article along with a letter, not-

---

[24]Minkowski to Hilbert, cited in [Rowe 2018a, 166].

ing connections with [Hilbert 1893], his fourteenth Paris problem [Hilbert 1900], and Steinitz's paper on abstract fields [Steinitz 1910]. She then added: "I have tried to deal exhaustively with the question of the rational representation of functions of an abstractly defined system by means of a basis (rational basis), and from there also to gain points of attack for the treatment of the finiteness problem. So new finiteness theorems have emerged; with conditions of a different kind than those that could be mastered up to now."[25] Emmy Noether's work on invariant theory soon caught the attention of others, as Hilbert began to spread the word. He and his colleagues eventually began to contemplate inviting her to habilitate in Göttingen, a plan that met with considerable resistance, as will be seen in the chapter that follows. By 1915, few mathematicians had any familiarity with classical invariant theory in the tradition of Noether's teacher, Paul Gordan. Hilbert's modern approach had found relatively few followers as well, in large part because he had abruptly left the field after publishing his final paper on the subject [Hilbert 1893]. Noether, on the other hand, had deep knowledge of both directions, which explains why the Göttingen mathematicians were willing to promote her candidacy for Habilitation, knowing full well that this would be a difficult struggle.

In closing this section, we cite the opinion of Colin McLarty, who offered this synopsis of how Noether's work stood in relation to that of her predecessors:

> ... Emmy Noether ... was in the most obvious sense a joint heir of Gordan and Hilbert. And she passionately sought to unify all mathematics in an algebraic axiomatic way. Corry has shown how Hilbert's axiomatics are never purely formal, nor even aim to found new subjects but always aim "to better define and understand existing mathematical and scientific theories" [Corry 2004, 161]. He aimed to organize classical subjects by paring each problem down to its stark essentials. For that very reason his axioms always have reference. They refer to the classical structures that motivate them. Gordan's algebra, on the other hand, was in his own terms "purely symbolic" so that "no meaning can be assigned to it." ... Noether's axiomatics combined the two. Her axioms create new subjects. They need not have classical referents. They are generally taken to have no specific referent, and sometimes understood to create new referents for themselves. But there is no use grappling with those conceptual ontological issues until we can make it as clear as one times one equals one how all of this is Mathematics. [McLarty 2012, 125]

---

[25]E. Noether to Hilbert, 4 May 1914, Nachlass Hilbert, SUB Göttingen.

3.5   Max Noether's Career in Retrospect

The mathematical world of Paul Gordan and Max Noether harkened back to the time of Alfred Clebsch, whose posthumous influence continued on through the journal he co-founded, *Mathematische Annalen*. Gordan and Noether both played important supporting roles with the *Annalen* [Rowe 2018b, 37–44], whereas Felix Klein assumed principal responsibility for the journal after Clebsch's death. However, by the 1890s Klein was intent on consolidating his power as Göttingen's senior mathematician. This placed him on a collision course with his old mathematical allies in Erlangen, both of whom strongly identified with the Clebschian tradition.

Emmy Noether had surely heard about the famous conflict between Gordan and Hilbert, whether from Ernst Fischer or directly from her father, and she likely knew that Klein had intervened to mediate in this dispute. As editor-in-chief of *Mathematische Annalen*, Klein increasingly took on the role of diplomat and visionary, leaving the routine work of editing the journal to others. Diplomacy in mathematical circles, however, can be a difficult matter, as Klein came to realize when he tried to forge new alliances while maintaining older ones. At the beginning of the new century, Klein wanted to place the *Annalen* firmly in the hands of his younger Göttingen colleague Hilbert, who had been a member of the editorial team since 1898. Klein had long been the driving force behind the journal, but he now proposed to resign from its editorial board and give Hilbert full authority to manage its affairs. He evidently hoped that the once sharp tensions between Gordan and Hilbert had by now subsided enough that no one in Erlangen would raise objections to this plan. Instead Klein received a letter from Max Noether pleading that he should reconsider:

> H[ilbert] is suitable for a rejuvenation, maybe even indispensable. But in matters concerning the DMV [German Mathematical Society] he has shown himself to be quite stubborn and one-sided, almost personal. You can perhaps judge him more accurately in this direction. Scientific interests alone can also affect a person's judgment, and some heads are never open to argument. I don't particularly like everything that he has included in recent issues of the *Annalen* .... – Under these circumstances [Gordan and I] would like to request that you remain on the editorial board for some time, and ... that you retain control of its management. G[ordan] even wishes for a kind of counterweight against an over-representation of work in Hilbert's direction. (Max Noether to Klein, 21 February 1901, cited in [Rowe 2018b, 43])

This private vote of no confidence in Hilbert's leadership qualities prompted Klein to change his plans. In fact, he would, retain his position as *de facto* head of *Mathematische Annalen* until shortly before his death in 1925. This anecdote clearly shows that Max Noether's opinions carried significant weight for Felix Klein.

Noether wrote his most important works in the 1870s and 1880s. He was considered to be the leading representative of algebraic geometry in Germany. Later he was referred to as *the last* major algebraic geometer in Germany, as after around 1890 Italian mathematicians dominated this field of research [Klein 1922, 7]. Together with Brill, Noether undertook the daunting task of writing a historical report on the development of the theory of algebraic functions for the German Mathematical Society (DMV) [Brill/Noether 1894]. Their joint work on this project took them three years to complete. Brill dealt with the older history up to Riemann, while Noether reported on subsequent developments. The DMV had originally enlisted Leopold Kronecker to write about the arithmetical direction of research, but this part of the report had to be dropped following Kronecker's unexpected death. Max Noether's most significant single achievement – often called his Fundamental Theorem or the $AF + BG$ theorem – was an essential tool for the Brill-Noether theory [Gray 2018, 256–257]. His daughter's account of that theory shows how its concepts can be rewritten in purely algebraic language by drawing on Dedekind's theory of ideals [Noether 1919, 197–201]. No doubt she discussed her own supplementary report with Max Noether before it appeared in print. One must imagine, too, that this was one of the first texts she advised young B.L. van der Waerden to read when they met in Göttingen in 1924 (see Section 5.2).

Max Noether celebrated the 50th jubilee of his doctorate in Erlangen on March 5, 1918. Emmy surely played a key role in making the arrangements for this event, a major milestone in her father's long career. Klein, of course, was among those who sent congratulations, to which Noether replied one week later:[26]

> I was pleased that you remembered the day of my 50th doctoral jubilee with heartfelt words. How everything during that space of time compresses together for me! Above all the decline and receding of directions that we considered indispensable, but also their passing over to other hands, in France and Italy. What remains, though, is our shared view of things, and that traces back to our strong point of departure with Clebsch. Such a bond is stronger than the effects of later influences; the influence [of Clebsch] was much stronger than I realized at the time. Here in Erlangen some very hard years are now behind me, but now that I have officially resigned from teaching the difficulties will perhaps gradually diminish. That our mutual relations have been revived again through the activities of my daughter has been a great source of satisfaction for me; I see every day how her creative powers grow and hope that these will lead to many new finds.[27]

Emmy surely had a clear picture of Max Noether's importance and reputation as an algebraic geometer, at the latest by 1908. With her new doctorate then in hand, she accompanied her parents on a trip to Rome, where they took

---

[26]Max Noether to Felix Klein, 13 March 1918, Nachlass Klein 12, SUB Göttingen.

[27]Indeed, just one day before, she sent Klein a postcard (Fig. 2.2) outlining the key ideas behind her classic paper [Noether 1918b].

part in the Fourth International Congress of Mathematicians. On this occasion, the Guccia medal was awarded to Francesco Severi. Corrado Segre presented the report on Severi's work in the name of the scientific committee, consisting of him, Henri Poincaré, and Max Noether. This medal was donated by Giovanni Guccia, the founder of the *Circolo Matematico di Palermo*, a prominent Italian mathematical society with some thousand members worldwide [Bongiorno/Curbera 2018]. Emmy Noether also became a member on this same occasion.

Just as Guccia cultivated international relations through the *Circolo Matematico*, so was Max Noether deeply interested in mathematics far beyond the borders of Germany. This is particularly evident from the scientific obituaries he published over many years in *Mathematische Annalen*. Except for the one written for his colleague Paul Gordan, all the others were dedicated to foreigners.[28] When Noether died in December 1921, Felix Klein, writing in the name of the editorial board, characterized this achievement in the following words: "[Noether] honored the memory of nine deceased mathematicians, masters with whom he was a kindred spirit, in extensive analytical-critical obituaries that form monuments, which together constitute an important contribution to the history of modern mathematics" [Klein 1922, 9].

Noether himself was so honored with an obituary in the *Annalen*, penned by Italy's three leading Italian algebraic geometers: Guido Castelnuovo, Federigo Enriques, and Severi [Castelnuovo/Enriques/Severi 1925]. For details about Max Noether's life, they relied on information from Emmy Noether. It was probably in connection with this obituary that these three Italians came up with the idea of presenting Emmy with a gift that surely meant a great deal to her. We only know about this, however, from a later source, a letter written by Emmy's colleague at Bryn Mawr College, Marguerite Lehr [Kimberling 1981, 56]. She recalled that on a wall in her office Emmy had hung a beautiful scroll. It depicted a metaphorical tree of algebraic geometry with many branches identified by various famous names. Prominent among them was a branch labeled "Max Noether." Emmy's father would not live to see his daughter's most creative work, which had only just begun when he died in 1921. But he did experience the satisfaction of witnessing her enormous first successes in Göttingen during the war years, crowned by her Habilitation in June 1919.

---

[28]Those so honored were the three British mathematicians Arthur Cayley, James Joseph Sylvester, and George Salmon; the two Italians Francesco Brioschi and Luigi Cremona; the Frenchman Charles Hermite; the Norwegian Sophus Lie; and the Dane Hieronymus Georg Zeuthen. Publication information on these obituaries and other scientific works by Max Noether can be found in [Castelnuovo/Enriques/Severi 1925].

Chapter 4

Emmy Noether's Long Struggle to Habilitate in Göttingen

4.1 Habilitation as the Last Hurdle

Seen from a long-term perspective, habilitation was a ritual of enormous significance for the German universities. As semi-autonomous institutions with close ties to church and state, these institutions maintained the right to confer academic titles but also to admit new members following age-old customs. In many ways, these traditional universities bore a striking likeness with craft guilds, and since several of them dated back to the late Middle Ages the resemblance was hardly fortuitous. From them emerged the research-oriented universities of the eighteenth and nineteenth centuries, one of the first being Göttingen's Georgia Augusta, founded in Hanover in 1734. Yet even these enlightened institutions, steeped as they were in a neohumanist ethos, tended to reinforce older patterns of training reminiscent of medieval guilds.

Neohumanism exalted the cultures of classical antiquity, a tradition reflected in the curriculum of the Gymnasien, which included heavy dosages of Latin and Greek. Since these elite schools for boys served for many decades as the exclusive pathway to entrance in the university system, those who taught at these institutions of higher learning took for granted that an educated person – such as those they encountered in their courses – possessed reasonable command, if not complete mastery, of these ancient languages. These capabilities in foreign languages were thus part of the standard repertoire of the educated class of citizens – the *Bildungsbürgertum* – which helps to account for the special status enjoyed by professors of classical philology, such as Göttingen's Ulrich von Wilamowitz-Moellendorff.

The young men who came to the university – most of whom were intent on pursuing careers outside of academia – entered a world utterly apart from the strictly disciplined atmosphere of the Gymnasium. Ideally at least, professors and

D. E. Rowe, M. Koreuber, *Proving It Her Way*, https://doi.org/10.1007/978-3-030-62811-6_4

students were expected to be motivated entirely by a spirit of higher learning enshrined in the twin principles of *Lehr- und Lernfreiheit*, the freedom to teach and to learn. There were no exams, no set requirements, not even a standard curriculum. Professors of mathematics, to be sure, taught introductory courses on subjects that every student was expected to know, though their content depended heavily on the tastes of the instructor. Teaching was understood as an expression of the personality of the *Dozent*, who performed more as a role model than as an instructor in the modern sense. The subject matter was mainly conveyed in courses called *Vorlesungen*, which in former times had been just that – a *Dozent* reading out loud from a text. This later evolved into a freer form of oral communication, which might be tied to a text, but usually was not. Students were expected not just to take notes, but to expand on them and, if possible, further develop the ideas set forth during lectures. They might also read textbook literature on the side, but the main focus was working over their lecture notes by producing an elaboration (*Ausarbeitung*) of the contents. Clearly, this system assumed a great deal of self-initiative and enthusiasm for learning by doing that had tremendous appeal for a certain type of intellectual elite. While free to explore new frontiers, students were expected to develop a deep sense of purpose and seriousness, an ethos that an older generation of scholars would then pass on to the next. This devotion to *Wissenschaft* (scholarship and science) emphasized the autonomy and self-sufficiency of the academic disciplines, ideals that had long made the inner sanctum of the universities rather similar to life in a religious order or a special type of guild.

The "apprentice-student" often studied at a number of universities before being promoted to a doctor of philosophy. As capstone for a course of study, he presented a doctoral thesis (*Inauguraldissertation*), which had to be successfully defended in a *Disputation*. This originally sufficed for habilitation, the procedure adopted for admitting new faculty members in the seventeenth century. By the nineteenth century, however, a second doctoral thesis was generally required. A post-doctoral candidate thus became a "journeyman-doctor" during this period of preparation for habilitation. To reach this first rung of the academic ladder, it was necessary to present the faculty with a finished "masterwork," the habilitation thesis (*Habilitationsschrift*). If this was found to be satisfactory. the candidate was awarded the *venia legendi*, which granted him the right to lecture as a *Privatdozent*, but without salary. He was merely entitled to collect the usual course fees from students, which were minimal. Since it was next to impossible for a *Privatdozent* to maintain himself without substantial private means, this system discouraged young scholars whose families were less wealthy from pursuing academic careers.

During the last decades of the nineteenth century, a bottleneck developed that left many in this state of limbo for many years. Hilbert had to wait six long years working as a *Privatdozent* in Königsberg before he was appointed associate professor there. Klein never had to face such a struggle: he acquired his doctorate at age 20, habilitated two years later in Göttingen, and was only 23 when he became full professor in Erlangen. In 1894, when he was dean of the Göttingen faculty, he engineered Hilbert's appointment within a matter of weeks. One of his colleagues criticized him for wanting to appoint an easygoing, amiable (bequem) younger man. To this Klein replied: "I want the most difficult of all" (Ich berufe mir den allerunbequemsten) [Blumenthal 1935, 399]; he knew how difficult Hilbert could be.

Temperamentally, Klein and Hilbert stood poles apart, though they seldom came into serious conflict with one another. Klein was the public face of Göttingen mathematics, the man who built its stellar staff and first gave the Georgia Augusta its central position in the modern mathematical world. He was also a universalist and an outspoken advocate for upholding the common interests within a unified philosophical faculty [Tobies 2019, 418–419]. On one occasion, he gave a much-publicized speech in hopes of garnering support for his position [Klein 1904], but relations between humanists and scientists in the faculty steadily worsened during the years afterward. Hilbert took little interest in the "tasks and future" of the faculty at large, particularly since he had more than enough to do in pursuing his own vast research agenda. This was anything but a lonely pursuit, and by 1900 his courses were beginning to attract many gifted students from all over the world. A decade later, Göttingen was literally swarming with talented doctoral and post-doctoral students in mathematics, supported by the largest teaching staff of any German university. Hilbert and his wife Käthe were on friendly terms with several Jewish families, and they were especially close with Hermann and Guste Minkowski. Their liberal views and lifestyle stood out in this small university town, whose guardians of civic virtue expected professors to appreciate that their social station demanded observance of traditional norms of behavior. Hilbert was having none of it, and with time his opposition to provincial attitudes became ever more evident and outspoken.

Klein and Hilbert were well aware that Emmy Noether's candidacy for *Habilitation* would face stiff opposition. Not that anyone was likely to challenge her general qualifications or mathematical competence, on the contrary. The issue at stake was actually very simple and straightforward: could a woman habilitate at one of the Prussian universities? In fact, this question had already been raised back in 1907 when a similar case arose at Bonn University. This concerned the zoologist Maria von Linden, who in 1908 gained a position at the new Institute of Parasitology, but without the right to teach since she lacked the *venia legendi*. Her case came up one year before the Prussian Ministry of Education decreed on August 18, 1908, that its universities would henceforth be open to qualified females.

Maria von Linden's petition to habilitate was forwarded to the Ministry, which decided to canvas the university faculties throughout Prussia as to whether the clause prohibiting women from *Habilitation* should be dropped from the regulations. The results of that survey showed that nearly all were opposed to making any such change. As a result, the Ministry decided not to change the *Habilitationsordnung*, a decision that spelled the end of von Linden's candidacy. Nevertheless, in 1910 she received a titular professorship as a researcher in Bonn [Tollmien 1990, 165].

Within the Göttingen philosophical faculty, however, several professors indicated that exemptions should be allowed in exceptional cases. In fact, nearly as many wanted to alter the regulation as those who supported upholding its prohibition of women. When a vote was taken, the latter group prevailed by only a single vote. Yet even more telling, particularly as a portent for the future, was the composition of these two opposing sides. For with the exception of the historian Max Lehmann, all those who supported the *possibility* of allowing women to be considered as candidates for habilitation were mathematicians or natural scientists, whereas those who were in the opposing camp were nearly all from the humanities.

After this vote was taken, the losing side submitted a minority report, composed by Hilbert, Runge, Lehmann, and the physicist Woldemar Voigt, and supported by Klein and Minkowski, the chemist Gustav Tammann, and the geophysicist Emil Wiechert. Its contents are less noteworthy than the response it elicited, in particular from the historian Karl Brandi, a proponent of the traditional, all-male university culture:

> ... I also believe that the minority report makes it necessary not only to emphasize that the previous scientific production of women in no way justifies making such a deep change in the character of the universities, but also to express once again that a great many of us in principle regard the entry of women into the organism of the universities as an impediment to the human and moral influence of the male university teacher on his hitherto largely homogeneous audience. I must confess that already in a mixed auditorium I feel a curb on the kind of complete lack of self-consciousness so necessary for our activity, so that I do not wish to dispense with the friendly tone of forthright expression and forthright trust. Our instruction should have a personal character and in my view, gender uniformity is required for this to be fully effective.

These views were certainly shared, in one form or another, by a large portion of the German professoriate, very few of whom had ever encountered the opposite sex in their own educational experience. A particularly noteworthy feature of this debate, however, was the way these sharp differences of opinion fell along disciplinary lines within the philosophical faculty. Not that this was unusual. Göttingen's humanists and scientists often found themselves on opposite sides of an argument, and in 1910 they agreed to go their separate ways by forming two separate departments: one for history and philology, the other for mathematics and natural sciences. Klein had long opposed such a division, but most felt that the divergence of interests between the two groups was too great to warrant maintaining a unified faculty. It soon transpired, however, that this arrangement left the underlying policy differences unresolved. The situation became even more intense after the outbreak of the Great War, which brought forth strong political antipathies as well. Under these circumstances, it became increasingly clear that many of the earlier disputes and conflicts between these factions were bound to continue.

4.2 Noether's Attempt to Habilitate

These earlier events were still very fresh in mind when Emmy Noether (Fig. 4.1) accepted the challenge of trying to habilitate in Göttingen. Klein realized, of course, that the university would have to seek an exemption from the Ministry of Education to evade the formal prohibition of women in the regulations. The first order of business, however, would be to convince reluctant colleagues in the philosophical faculty that Noether posed no threat to the status quo. To what extent she was made aware of the circumstances remains unclear, but in any event on 20 July 1915 she submitted her application to the Mathematics and Natural Sciences Department of the Philosophical Faculty of the University of Göttingen, along with the required fee of 100 Marks. Exactly one week earlier, she had delivered a lecture in the Mathematical Society on finiteness questions in invariant theory, a performance that presumably convinced any remaining doubters among those who heard her speak. As her habilitation thesis, she submitted the paper "Fields and Systems of Rational Functions" [Noether 1915], which she had sent to Hilbert one year before.

On the very day Noether submitted her application, Edmund Landau, who was then head of the department as well as dean of the faculty, called a preparatory meeting to discuss how to proceed. As usual, a commission was formed, some of whose members agreed to write reports evaluating Noether's abilities as well as her suitability for membership in the faculty. This commission consisted of: the four mathematicians Landau, Klein, Hilbert and Constantin Carathéodory; the applied mathematician Carl Runge; the physicists Woldemar Voigt and Peter Debye; and the astronomer Johannes Hartmann. Because of the larger implications of Noether's case, those present also decided to invite the Egyptologist Kurt Sethe, who was head of the historical-philological department at the time, to be a member of the commission as well. Sethe accepted, but he took no active part in any deliberations in accordance with a general understanding that matters concerning habilitation proceedings were to be handled exclusively by the responsible department. The four mathematicians in the commission agreed to submit reports, which served as the basis for discussion when the members reconvened in October.

One week after this meeting, Klein wrote to the Ministry of Education with some personal observations about the prospective candidate. Recalling the one semester she had spent in Göttingen over a decade before, he then thought that she was attempting something she could not attain. What changed his mind was a lengthy visit two years before, when Emmy accompanied her father, who wanted to gather information from both her and Klein in preparation for writing his obituary of Paul Gordan [M. Noether 1914]. "I saw to my surprise," wrote Klein,

> that she not only had a full command of one of my earlier research fields, the theory of quintic equations, but that she could inform me about several details that were new and gave me great satisfaction. Since then I am convinced that Miss Noether fulfills by all means the conditions that we customarily require of those who apply to habilitate, indeed that she is <u>superior in quality to the average candidates</u> that we have admitted in recent years.[1]

Among the reports offered by the four mathematicians, one finds significant divergences of opinion, although all reached the conclusion that the committee should recommend to the department the approval of Noether's application. Only Hilbert, who was appointed principal reporter (erster Gutachter), treated Noether's candidacy in a gender-free fashion, in accordance with his conviction that scientific abilities and accomplishments were the only relevant criteria. Under normal circumstances, the philosophical faculty would have routinely rubber-stamped this decision, passed it on to the Kurator, who might have added a few brief remarks about the candidate before sending the application to the Prussian Ministry of Education for final approval. In short, the rest would have been a formality. All members of the commission were acutely aware, though, that Noether's case was anything but normal. Indeed, the mathematicians' initiative raised a general issue that had been debated earlier in the Göttingen philosophical faculty, which remained sharply divided over whether a woman could under any conditions be allowed to habilitate.

Many of the commission members had mixed feelings about this as well. In contrast to Hilbert, Landau's report defended the traditional conception of an academic faculty as a body that needed to scrutinize the personal qualities of applicants for membership with an eye toward ensuring harmonious cooperation. In this regard, he emphasized that a number of male applicants had been rejected as unsuitable for personal reasons. Landau thus conceded that the highly subjective quality of "suitability" had to be recognized as an important factor, though his report contained no direct indications why this was relevant in Noether's case. Indeed, since his concrete remarks regarding her character were entirely positive, one can safely conclude that, for Landau, *her sex was the only real issue* at stake. "How easy this decision would be for us if this were a man with exactly these accomplishments, skills as a lecturer, and earnest ambitions. I would much prefer it, if this extension of our teaching program could be accomplished without the habilitation of a <u>lady</u>" [Tollmien 1990, 176].

In line with the consensus view, Landau noted Noether's highly exceptional mathematical abilities, though he did so by drawing on his uniformly negative experiences with female students, especially when it came to their productive

---

[1]Klein to Ministry, 27 July 1915, Nachlass Klein 2G, SUB Göttingen.

capacities. "I think the female brain is unsuitable for mathematical production," he wrote, "but I regard Miss N[oether] as one of the rare exceptions" (*ibid.*). Landau did not consider her to be a genius, not even a first-rate scholar, and then went so far as to predict that she would be accorded much unjustified fame, just as had been the case with Sonya Kovalevskaya. Still, he could not begrudge giving her what she was due, namely the chance to habilitate.

Landau's report was the least enthusiastic of the four, mainly because his general hostility to female mathematicians strongly colored the views he expressed in it. Yet even Klein took a rather reserved view of the matter, stating at the outset of his report that he had no wish to advocate women's rights to pursue academic careers. He stressed that Noether's case was entirely exceptional, distancing himself from "those who generally recommend the extension of academic studies to women," including opening the doors to habilitation. Instead, Klein argued simply from the point of view of self-interest: Noether was an extraordinary talent, so acquiring her services would serve to strengthen mathematics in Göttingen. In support of this, he noted that she was better qualified than the average candidate accepted by the faculty in recent years [Tollmien 1990, 175].

Carathéodory's report, which was likely written after conferring with Klein, took essentially the same position. He affirmed Emmy Noether's special qualifications, while emphasizing her knowledge of invariant theory and her background as a student of Max Noether and Paul Gordan. He further pointed out that classical invariant theory had fallen out of favor after Hilbert managed to solve its central problem in one fell swoop (see Chapter 3). The result was that the generation to which he and his colleague Landau belonged had little idea of the subject, which for many years was unjustifiably devalued. Alluding to Hermann Minkowski's mathematical formalization of Einstein's special theory of relativity, Carathéodory wrote: "... now the time seems to have come [...] when invariant theory is called upon to form the basis for the youngest and deepest physical theories." As a former student of Minkowski, it should come as no surprise that Carathéodory emphasized this connection in describing Emmy Noether as "someone who has picked up the broken thread again after 20 years and not only understood the existing theory in the best sense of the word, but who has also added something new and valuable." This, he added, was no mere coincidence, since she received her training in Erlangen, one of the last places where classical invariant theory was still cultivated.[2] Largely echoing Klein's argument, Carathéodory concluded:

> It is, in my opinion, <u>not</u> to be assumed that there is anyone else in the world today, who would be obtainable and who could replace Miss Noether. As the matter stands, Miss Noether is a <u>singular phenomenon</u>,

---

[2]As further indication of this trend, one can point to the career of Eduard Study who, unlike Hilbert, continued to work on invariant theory. Remaining copies of Study's important book, *Methoden zur Theorie der ternären Formen*, first published in 1889 by Teubner, had to be discarded by the publisher for lack of buyers. Decades later, when classical invariant theory enjoyed a true Renaissance, Springer produced a new edition in 1982 [Brieskorn/Purkert 2018, 708].

who can be beneficial for the further scientific development of the university, and this is the main reason why, disregarding everything else, I am in favor of opening this habilitation procedure. [Tollmien 1990, 177]

Figure 4.1: Emmy Noether, ca. 1920 (SUB Göttingen, Sammlung Voit, no. 4)

Unlike the other three reporters, who only had a general impression of Noether's work, Hilbert had studied several of her papers carefully. So he was uniquely qualified to offer a competent judgment of her abilities, not least because her most recent publications represented a continuation of his own earlier researches. He thus characterized her habilitation thesis [Noether 1915] as "the successful execution of a part of the large program that I set forth with regard to finiteness questions" (*ibid.*). He also wrote about his delight that she had been able to prove a result he had recently conjectured concerning the finiteness of a system of infinitely many basis forms. Finally, Hilbert emphasized her versatility in applying formal theoretical methods, as demonstrated in her recently published paper [Noether 1916].

As the summer vacation loomed, Noether's Habilitation-Commission only met again on 29 October, when the winter semester 1915/16 was underway. Since all four reports praised the candidates qualifications and recommended that the department request an exemption from the exclusion clause in the habilitation regulations, there was presumably little to debate at this point in the procedure. Landau moved that the commission recommend approval of Noether's application, and all members with the exception of the astronomer Johannes Hartmann voted in favor. After this, the matter was taken up by the mathematics-natural sciences department. During the course of its deliberations, it became apparent that others shared Hartmann's reservations and opposed the commission's findings. Whereas proponents stressed the exceptional nature of Noether's case, opponents argued that it would establish an undesirable precedent. When the final vote was taken, the motion to accept the commission's recommendations passed 10 to 7 with 2 abstentions [Tollmien 1990, 171]. Afterward, some of those who voted with the losing side announced they would file a minority report. Arrangements were also made for Emmy Noether to deliver a public lecture before the Mathematical Society, which interested members of the philosophical faculty were invited to attend. Although largely a formality, this lecture provided an opportunity for the faculty to judge whether she possessed the requisite teaching skills to be appointed as a private lecturer. Three days later, on 9 November 1915, Noether spoke on transcendental integers, the topic of her paper [Noether 1916]. Afterward, she wrote to Ernst Fischer: "Even our geographer came to hear it and found it rather too abstract; the faculty wants to make sure it's not going to be duped at the meeting by the mathematicians" [Dick 1970/1981, 31–32].

The results from the earlier meeting on November 6 were also communicated to the historical-philological department, which met on 10 November to discuss their implications. Afterward, the head of the department announced that this was a matter of urgent concern to the entire faculty, "... in view of the fundamental importance of the present case, which would be a complete novelty of the greatest importance for German university life." Joint deliberations of its two sections were particularly justified in view of the fact that, while "the general responsibility in habilitation matters lies with the individual departments," some members of the mathematics-natural sciences department had themselves expressed the wish for a joint meeting [Tollmien 1990, 171].

This meeting of the entire faculty then convened on 18 November in a tense atmosphere. Edmund Landau presided as dean of the faculty. Although minutes from this meeting have not survived, it clearly turned into an explosive session, during which Hilbert apparently lost his temper. At one point, he scolded the humanists on the other side, an incident remembered by his perhaps legendary remark: "gentlemen, we're in a university and not a bathing establishment" ("meine Herren, wir befinden uns an einer Universität und nicht in einer Badeanstalt").[3]

Whether or not he actually uttered these words, the discussion must have become unusually heated, as Hilbert definitely did criticize two members of the other department for their inability to focus on the merits of Noether's case. This was documented by Cordula Tollmien, who reproduced a draft of Hilbert's letter of apology to these two colleagues, the classical philologists Richard Reitzenstein and

---

[3]There appears to be no credible source asserting that Hilbert made this remark, although Hermann Weyl treated this as an established fact. In his memorial address for Emmy Noether, he speculated further that "probably he provoked the adversaries [the humanists in the faculty] even more by that remark" [Weyl 1935, 431].

Max Pohlenz, both of whom had filed a written complaint with Landau, in which they charged Hilbert with uncollegial behavior [Tollmien 1990, 178]. Evidently, Hilbert's attack on them had included remarks, which they interpreted as an invitation to leave the meeting. He denied any intention "to personally insult any of my colleagues" or to expel any faculty member. What he meant to express was the view that university professors should concern themselves with scientific matters alone, leaving social and political pursuits to the world outside. Regarding collegiality, Hilbert drew a line when it came to "protecting real interests."

In his draft of this communication, however, Hilbert struck out a highly significant passage that read:

> In order to prevent misunderstandings, I must add that it was my full intention to say that in the past 20 years the historical-philological division, under the leadership of the classical and Germanic philologists, has at every opportunity ([matters concerning] matriculation, foreigners, women, doctoral candidates, habilitations) tried to thwart by all means possible my efforts, which are solely aimed at promoting science. Even after the division of the faculty and the changes in personnel, I still thought I had to assume that this attitude persisted, and the reason for it is to be found in the vast and, I believe, insurmountable gulf that separates my view of the professor's responsibility to science and that of my opponents.

Hilbert's outburst at the meeting on 18 November stemmed from his strongly held conviction that Noether's qualifications should have been the only relevant issue, a view most of his allies clearly did not share, as revealed by the three other reports that Landau, Klein, and Carathéodory prepared for the habilitation commission. As for the outcome of the meeting, this hinged on the results of two motions put before the faculty. First, a preliminary vote was taken on the question: "who opposes the admission of a woman to habilitate under all circumstances?" The outcome – 17 to 14 with one abstention – clearly revealed the depth of the opposition. Since the scientists outnumbered the humanists by 19 to 13, this vote demonstrated that the astronomer Hartmann was hardly a lone voice of dissent: in total, seven of the 19 scientists categorically opposed extending the right of habilitation to women [Tollmien 1990, 172]. The faculty then voted on a second motion, which recommended that the Minister of Education (Kultusminister) reject the initiative taken by the mathematics-natural sciences department. Since this motion clearly represented direct interference of one department in the affairs of another, three supporters of the first motion decided to abstain during this second round of voting. The result was thus 14 votes for and against, which placed the outcome in Landau's hands as dean. As a result, this resolution was defeated by the slimmest of margins.

Hilbert clearly felt vindicated by this turn of events. In a letter from his department to the Ministry from 4 December 1915, signed by Hilbert, he noted that the faculty as a whole had defeated a motion recommending that the Minister refuse to allow Noether's habilitation and, furthermore, that the faculty's philological- historical department also abandoned its original plan to submit a counter-presentation [Tollmien 1990, 179]. Having survived this confrontation, the mathematics-natural sciences department now had a free hand to pursue this matter and did so by convening on the very same day. The five mathematicians – Landau, Klein, Hilbert, Carathéodory, and Runge – formulated a petition that would go to the Ministry, and this draft was approved by a vote of 10 to 6 with one abstention.

The following day, the six who lost that vote filed a minority report. Their main argument against Noether's candidacy was simply the fear that it would mark the beginning of a domino effect:

> ... it cannot in any way be denied that with the admission of this first woman the question of whether women are allowed to habilitate at all would be answered in the affirmative. This would open up a new life path for numerous female students and the scientific level at the German universities would undoubtedly fall as a result of this increasing feminization. All members of the faculty agree – and those in the majority expressly concede this – that only in the most exceptional cases can a female head produce creative scientific achievements.
>
> But a woman, in particular, is altogether unsuitable for regular instruction of our students because of the phenomena connected with the female organism. With the admission to habilitation would, in principle, also follow admission of women to the next stages of an academic career, to professorships, and consequently membership in the faculty and the senate. For it would be an obvious hardship if women were permitted to begin an academic career, but were then prevented from progressing afterward. [Tollmien 1990, 173]

One hardly needs to read between the lines to recognize the entirely self-serving nature of this document. If nothing else, however, it fully confirms that those who supported it simply wished to preserve the university as an exclusively male sanctum. That being so, the logic behind its argumentation was unassailable: even if habilitation was by no means the final barrier women would encounter, it was the decisive one.

As a kind of postscript, the astronomer Johannes Hartmann added quite another kind argument that brought the situation in Germany in November 1915 into clear focus. His considerations touched on the role of women in social and family life as well as the future lives of the young men presently serving their country in the Great War.

> If Germany is now able to successfully face a world of enemies, we owe it in large part to our German women and mothers, who have brought up large numbers of sons. Every measure that furthers the equality of women and facilitates their independent attitudes and lifestyles brings

with it certain dangers for family life, for fulfilling those tasks from which women cannot be relieved but which to some women, once they have turned their interest to scientific work, might seem uncomfortable. In particular, a regular academic teaching position can hardly be reconciled with the tasks of a married woman. In the interest of our offspring, it would surely be undesirable if mentally superior women were more and more removed from family life. – Quite apart from this general consideration, the current point in time would seem particularly unsuitable for taking a step with such far-reaching consequences. While Germany's sons must carry out the bloody business of war far from their home place of work, we have in many instances necessarily utilized women to replace the missing male workforce. As welcome as this kind of help is, it would be reprehensible if a whole great profession, which till now was exclusively practiced by men, were given over to women without any need. Our private lecturers returning from the field, each of whom sacrificed a greater or lesser part of their health for the fatherland, would surely greet such competition, which had emerged during their absence, with very mixed feelings. Precisely because academic teaching is one of the few professions which a man with a war injury can practice almost without any difficulties, we should expect that some students returning from the field will want to pursue this career. It would be irresponsible and deeply troubling if in this very profession their prospects were made more difficult due to competition from women. [Tollmien 1990, 173–174]

The authors of the department's petition had ample time to read and react both to the minority report as well as Hartmann's emotionally charged addendum to it. They took pains to explain that in requesting an exemption for Emmy Noether from the prohibition of women in the decree of 1908 they were in no sense contesting the legitimacy of that stipulation. They noted, however, that it was issued shortly before women had been allowed to matriculate. Nevertheless, the burden of their argument rested on the contention that Noether's talents were so exceptional as to make her case unique and, hence, in no way would it constitute grounds for subsequent applications. Moreover, as further substantiation of the latter claim, they underscored that this matter had not come about through her own initiative but rather she had been encouraged to apply for habilitation by mathematicians on the faculty. Their encouragement was apparently formalized after she delivered her lecture on November 9, which was deemed "pedagogically successful." With regard to the minority report, the majority merely noted that its signers recognized Noether's scientific qualifications, which were examined by a special commission, so their dissent was based solely on general reluctance to admit a woman. The majority further affirmed that "Miss Noether's achievements exceed the average level of private lecturers in mathematics who had previously been accepted in Göttingen."

They then turned to Hartmann's arguments, particularly his contention that allowing a woman to habilitate in wartime amounted to *de facto* discrimination against war veterans. The majority asserted that, on the contrary, Noether's candidacy "poses no threat to men returning from the field or to future private lecturers in mathematics." First, the department had never instituted a *numerus clausus* for lecturers in mathematics. Second, an effort had been made even before the war to recruit new lecturers in mathematics, but this search had failed to turn up even one candidate who met the department's high standards. Third, although Miss Noether would only be covering a part of the present gap in course offerings, there was little chance that another woman would be recruited any time soon. The reason for this – as noted in the minority report – was the shared opinion that "only in exceptional cases can a female head be creatively productive in mathematics, let alone display Miss Noether's achievements" [Tollmien 1990, 164].

Hilbert probably anticipated that the Ministry would not go along with the mathematicians' request. As a precaution, he therefore signaled his willingness to compromise in the letter from December 4 cited above. This was written just after he and Einstein had ended a flurry of correspondence in November. During this time, Einstein published four famous notes that culminated with his new gravitational field equations, whereas Hilbert submitted his first note on general relativity [Hilbert 1915]. In his letter, Hilbert briefly alluded to these fast-breaking developments, adding that "here I have Miss Emmy Noether as my most successful assistant." He then went on to say that if the Minister should be disinclined to approve her habilitation, he would like to have the opportunity to discuss the situation personally [Tollmien 1990, 179]. This communication bypassed Göttingen's Kurator, Ernst Osterrath, the official responsible for transmitting faculty decisions to the Ministry. In the past, Klein and Hilbert had often taken up direct negotiations with the Ministry, in keeping with Friedrich Althoff's administrative style as head of the Prussian university system. Hilbert was also surely aware that the Kurator opposed allowing women to habilitate. He thus had reason to be concerned that his negative opinion might tip the scales against Noether.

On 9 December, Osterrath submitted the department's petition as well as the other related documents to the Ministry along with brief remarks indicating why he could not support this proposal. Rather surprisingly, he named Klein rather than Hilbert as its prime initiator. Doing so surely lent *gravitas* to the whole matter, since Klein was unquestionably the leading spokesman for all matters touching on mathematics education in Prussia. Since 1908 he represented Göttingen University in the Prussian House of Lords (*Herrenhaus*), where he occasionally gave speeches on educational matters. Osterrath was already appointed Kurator before the division of the philosophical faculty in 1908, so he knew very well that Hilbert was a dogged fighter whose various liberal causes rankled opponents in the faculty. He likely also knew that Emmy Noether's case was dear to Klein's heart, not least due to his long friendship with her father. As Kurator, he characterized Klein's position as generally opposed to the habilitation of women. However, he added,

"[Klein] believes that Miss Noether is so exceptionally competent that an exception can be justified," a view Osterrath found unconvincing. He instead supported the views expressed in the minority report as well as the reasoning in the decree of 1908, which remained in his eyes just as valid as before.

What happened afterward remains somewhat unclear owing to lack of documentary evidence. In any event, the Prussian Ministry of Education must have made clear that it would not approve Noether's habilitation at this time. Nevertheless, the meeting Hilbert had requested did take place, and he easily reached an understanding with the Ministry's Director of University Affairs, Otto Naumann. The faculty's petition was set aside, but Emmy Noether would be allowed to teach courses offered under Hilbert's name. She did so regularly, starting in the winter semester of 1916/17 when she taught a two-hour seminar on invariant theory for mathematicians and physicists, a topic of burning interest in Göttingen.

In the summer of 1917, members of the Göttingen faculty learned that Frankfurt University had taken an interest in Emmy Noether's case. Fearing that she might be lured away from the Georgia August, they inquired whether it might be possible to reactivate the earlier petition, which had never been officially answered. Only days later, they received an emphatic answer from Otto Naumann, who simply asserted that there was no chance Noether would be able to habilitate in Frankfurt or anywhere else in Prussia.

> The same regulations apply to the University of Frankfurt as to the other universities regarding the admission of women to the teaching profession; i.e., they are not allowed to become private lecturers. It is also quite impossible to make an exception for one university. Thus, your fear that Miss Noether could go to Frankfurt to obtain the venia legendi is unfounded; she will not be admitted there any more than in Göttingen or at another university. The Minister of Education has repeatedly stated that he firmly holds to his predecessor's stipulation that women should not be allowed to teach at universities. In any case, you will not lose Miss Noether as a private lecturer to the University of Frankfurt. [Tollmien 1990, 180]

Not long afterward, Friedrich Schmidt-Ott became Minister of Education. In November 1917, Schmidt-Ott wrote to Göttingen's Kurator, evidently the first official response to the petition on behalf of Emmy Noether from two years earlier. He took a neutral position on the general question of allowing women to habilitate, declaring that nothing could change until such time as the faculties themselves voiced a desire to see a liberalization of the regulations:

> The admission of women to habilitate as private lecturers continues to face considerable misgivings in academic circles. Since this question can only be decided as a matter of principle, I am unable to approve any exceptions, even if this means that certain hardships are unavoidable in individual cases. Should the general opinion of the faculties change

from those then taken into account in the decree of May 29, 1908, I would be willing to consider the question again. [Tollmien 1990, 181]

The professors' opinions, by and large, did not change, but the world they lived in certainly did. Chancellor Bethmann Hollweg's fall ushered in what amounted to a military dictatorship, led by Chief of the General Staff Paul von Hindenburg and his deputy, General Erich Ludendorff. Meanwhile, the Kaiser retreated more and more from view, leaving all meaningful decision-making to the German High Command. As the military situation deteriorated, pressure mounted for Wilhelm II to abdicate, a step he took most reluctantly on November 9, 1918, fleeing to Holland the following day. His departure spelled the end of the constitutional monarchies in Germany, as one by one the kings and princes of the German states willingly abdicated.

Meanwhile, mutiny and chaos reigned, as many thought Germany was ripe for the revolution Karl Marx had long before predicted. The long-suppressed Social Democrats quickly tried to seize power, but internal dissension and fears of Bolshevist influences slowed these efforts. The Social Democratic Party of Germany (SPD) had long identified with Marxism, but with the outbreak of the Great War the party subordinated international solidarity of the working class to the pressing nationalist interests of the moment. As the fighting dragged on, dissension within its ranks grew, and in 1917 the party split, forming the Majority Social Democrats (MSPD), who supported the war, and the rival Independent Social Democrats (USPD), many of whom had strong pacifist leanings.

Two years later, Emmy Noether joined the small, but growing Göttingen USPD, which over the course of that year tripled its membership from 146 to 450. The following year in the June elections, the USPD won 2,500 votes in Göttingen, second only to the MSPD, which gained 4,500 votes [McLarty 2005, 438]. Noether remained in the party until 1922, when many in the USPD rejoined the Majority Social Democrats. Two years later, she dropped her membership in the SPD, perhaps as a result of the short-lived stabilization of political life in Germany during the Weimar Republic. Various hints as to the depth of her leftist politics were voiced at the time of her death, but virtually no sources remain to illuminate what drew her into the political arena or caused her to leave it later (see [McLarty 2005] for a discussion of the contrasting views of Pavel Alexandrov and Hermann Weyl). To what degree was she politically active? And, if so, when and with what aims? All that can be said with any assurance is that Emmy Noether was branded a "Marxist" by many in Göttingen, a label that would have been applicable to any member of the intelligentsia who voted for the SPD.

In any event, very few German academics sympathized with the MSPD and even fewer identified with the USPD. Still, Noether's politics placed her in unusually interesting company. Einstein was elated by the collapse of the monarchy, and he too sympathized with figures on the left, such as the USPD politician Kurt Eisner, who was assassinated in Munich in February 1919. Probably Einstein had no idea of Noether's political leanings, but soon after the war he wrote an emphatic letter to Klein pleading that something be done about her case. He even intimated that if the Göttingen faculty failed to act, he would intercede himself: "After receiving the new paper ["Invariante Variationsprobleme" [Noether 1918b]] by Frl. Noether I again felt that it is a great injustice that she is denied the *venia legendi*. I would very much favor that we take energetic steps with the Ministry. If you do not regard this as possible, then I will make the efforts myself" (Einstein to Klein, 27 December 1918, [Einstein 1998, 976]).

Klein needed no prodding, and he later gave the following assessment of Emmy Noether's most recent accomplishments: "In the past year she has completed a number of theoretical studies that are superior to the achievements made by all others in the same period (including the work of the full professors) ...." Still, Einstein's support came at an opportune time. The Ministry of Education was now headed by two prominent socialists: Adolph Hoffmann (USPD) and Konrad Haenisch (SPD). By the end of January the mathematics-natural sciences department voted to approve Noether's petition to habilitate, and by mid-February the university communicated the same to the Ministry, this time with the approval of the entire philosophical faculty:

> The changed political conditions, which have led to a comprehensive expansion of women's rights, have given our mathematicians the hope that an application in this direction would now be successful. At a meeting held on 31 January 1919, Frl. Dr. Emmy Noether's wish to renew her application for habilitation, which had been rejected at the time, was endorsed by an overwhelming majority in the Department of Mathematics and Natural Sciences, which therefore renews her application from November 26, 1915 to be allowed to habilitate as an exceptional case. The department does not wish to prompt a general decision with regard to the admissibility of women in the teaching profession, but rather bases its application, as before, on the applicant's extraordinarily high level of mathematical talent and academic performance. ... [Tollmien 1990, 183–184]

Much had changed, indeed, as women in Germany had now gained the right to vote, which they exercised in the election of 19 January 1919 for the Weimar National Assembly, the constitutional convention for the new Republic. Still, the stricture in the decree of 1908 that prevented women from habilitating had yet to be lifted when Noether applied for an exemption. Minister Haenisch (SPD) surely saw no reason not to do so, and on 4 June 1919, after delivering her trial lecture on "Questions of Module Theory," Noether was finally given the *venia legendi* entitling her to teach courses as a Privatdozent. One year later, the philosopher Edith Stein, who had been denied the opportunity to habilitate on more than one occasion, presented a petition to the Ministry arguing that discrimination by gender should be disallowed in future habilitation proceedings. Haenisch agreed, thereby overturning the prohibitive provision in the decree from 1908.

Chapter 5

Noether's Early Contributions to Modern Algebra

5.1 On the Rise of Abstract Algebra

As described in preceding chapters, Noether's work on invariant theory broke new ground that led the Göttingen mathematicians, but first and foremost Hilbert, to invite her to habilitate there. At the time, no one would have imagined that her expertise in this field would prove decisive for clarifying how so-called conservation laws are related to symmetries of group actions that leave a variational system invariant. This caused a real stir of interest in 1918, but soon thereafter the Noether theorems were largely forgotten by the main actors, Einstein, Weyl, et al., even though they continued to pursue schemes for uniting gravity and electromagnetism into a single field theory. In the meantime, however, Noether's earlier work on invariant theory was by no means overlooked.

This chapter attempts to provide a glimpse of the many facets of Noether's school in abstract algebra, one of the most famous success stories in the history of modern mathematics. Noether supervised several doctoral students, of course, some of whom made important contributions to modern algebra. Yet her influence extended far beyond the circle of those who took their doctorates under her, as illustrated by four mathematicians whose careers were closely linked with Emmy Noether's: B.L. van der Waerden, Pavel Alexandrov, Helmut Hasse, and Olga Taussky. All four came from distinctively different mathematical cultures nothing like Noether's world in Göttingen, and yet they naturally gravitated into her school and its broader network, which was itself part of a larger cultural movement in early twentieth-century mathematics.

Today Emmy Noether holds a secure place in the history of algebra, as her work represents not only the culmination of a long-standing development but also

© The Author(s), under exclusive license to Springer Nature Switzerland AG 2020
D. E. Rowe, M. Koreuber, *Proving It Her Way*, https://doi.org/10.1007/978-3-030-62811-6_5

a fresh beginning for a new form of abstract mathematics.[1] Reflecting on this, Israel Kleiner wrote:

> ... Noether was not the only, nor even the only major contributor to the abstract, axiomatic approach to algebra. Among her predecessors who contributed to the genre were Cayley and Frobenius in group theory, Dedekind in lattice theory, Weber and Steinitz in field theory, and Wedderburn and Dickson in the theory of hypercomplex systems. Among her contemporaries, Albert in the U.S. and Artin in Germany stand out. ... [2]
>
> ... [Noether] dealt with just about the whole range of subject matter of the algebraic tradition of the nineteenth and early twentieth centuries (with the possible exception of group theory proper). What is significant is that she transformed that subject matter, thereby originating a new algebraic tradition – what has come to be known as modern or abstract algebra. [Kleiner 2007, 91–92]

In short, Noether helped to spearhead a broad cultural movement,[3] and her influence within this dynamic sphere of activity was truly pervasive. B.L. van der Waerden summed up her mathematics as flowing from a single unified principle that served as her maxim:

> All relations between numbers, functions, and operations become perspicuous, capable of generalization, and truly fruitful, when they are detached from specific examples and traced back to conceptual connections. [van der Waerden 1935, 469]

5.2   Van der Waerden in Göttingen

B.L. van der Waerden's father, Theo, was a civil engineer and he also taught mathematics in Amsterdam, though he apparently made no effort to push his son in that direction. Nor did he need to, since his son was a born mathematician who learned all facets of the subject almost effortlessly. From the beginning, his scientific interests were extremely broad. At the university, he took courses offered by the philosopher-mathematician Gerrit Mannoury (a close friend of his father), the invariant-theorist Roland Weitzenböck, the geometer Hendrik de Vries, and the famous topologist, L.E.J. Brouwer. By this time, however, Brouwer was no longer working on topology but rather pursuing his first love, a new approach to the foundations of mathematics known as intuitionism.

During five years of study in Amsterdam, van der Waerden also learned a good deal of classical algebra, beginning with a course he took with de Vries.

---

[1] For a recent survey of the subject, see [Gray 2018].

[2] On the Chicago tradition of Dickson and Albert, see [Fenster 2007].

[3] The role of the Noether school in this cultural movement is described in [Koreuber 2015, Kap. 4].

For an aspiring young mathematician in the early 1920s, van der Waerden surely had acquired a solid background in Amsterdam, and yet he soon learned that marvelous new wholly unsuspected things awaited him in Göttingen. He later recalled his "apprentice years" there, including his first encounters with Noether.

> Brouwer recommended that I particularly stick with Emmy Noether. At that time he was quite hostile toward Hilbert due to the argument over intuitionism versus formalism.[4] But Brouwer thought highly of the younger mathematicians in Göttingen, such as Emmy Noether, Ostrowski and Hellmuth Kneser. So I attended Emmy Noether's lectures and soon met her personally. She was a very peculiar personality, roughly built with a large nose and inelegant movements, and she trudged while lecturing, sometimes she crushed a piece of chalk that had broken off ... the opposite of an elegant lady. As Hermann Weyl expressed it in his obituary: "The graces were not at her cradle." But these were outward appearances. More importantly, she was an altogether good person, free of all selfishness, free of all vanity, never posing and always willing to help everyone whenever she could. Her lectures were not well polished. She presented what she had just been thinking about, and she tried to improve the presentation during the lecture. It went like this: even before she had finished formulating a theorem, she quickly brought a better formulation. That, of course, did not make understanding any easier, on the contrary. But if you listened carefully and tried to think along, you could learn more than from a perfectly polished lecture. [van der Waerden 1997]

Brouwer stood on very good terms with Emmy Noether, who was nearly his age. They had been acquainted for some time, having first met one another in 1911 during the DMV conference held in Karlsruhe.[5] Brouwer also knew the far younger Hellmuth Kneser, then an assistant of Richard Courant. This particular connection would also prove to be of great value for van der Waerden. On 21 October 1924, Brouwer wrote Kneser to to inform him of the latter's imminent arrival:

> In a few days, a student of mine (or actually rather of Weitzenböck's) will come to Göttingen for the winter term. His name is van der Waerden; he is very bright and has already published things (especially about invariant theory). I do not know whether the formalities a foreigner has to go through in order to register at the University are difficult at the moment; at any rate it would be very valuable for van der Waerden if he could find help and guidance. May he then contact you?[6]

---

[4]On intuitionism in the 1920s, see [Hesseling 2003].

[5]She recalled that meeting in a postcard sent to Brouwer, 7 October 1919, Brouwer Papers, Noord-Hollands Archief, Haarlem.

[6]Brouwer to Kneser, 21 October 1924, Nachlass Kneser, SUB Göttingen.

Soon afterward, van der Waerden began meeting Kneser regularly for lunch, after which they would often take strolls through the woods just outside the town. Many years later, van der Waerden recalled how Kneser would

> ... start on a certain subject and make a few remarks which I couldn't understand at all. Then I would say to him that I would like to learn about that subject. Where could I find out about it? So he would give me the names of some books which I could find in the Lesezimmer. A day or so later I would be able to answer his questions and also make some significant remarks of my own, and then I learned much more. [Reid 1970, 162]

In 1924, at age 26, Kneser delivered a lecture on combinatorial topology at the annual meeting of the German Mathematical Society in Innsbruck [Kneser 1926]. While summarizing the current state of knowledge, he proceeded to sketch a program for future research that would only reach its zenith in the 1960s.[7] One of its major objectives was to prove what Kneser dubbed the *Hauptvermutung* (principal conjecture).[8] Five years later, van der Waerden presented an overview of research on combinatorial topology at the 1929 DMV conference in Prague [van der Waerden 1930a], at which Kneser was present. Van der Waerden began by referring back to Kneser's Innsbruck lecture before proceeding to describe more recent work. Emmy Noether reported briefly about this lecture in a letter to Pavel Alexandrov:

> Prague showed that there is great interest in topology. In no lecture were there as many auditors as those who heard v. d. Waerden's report. It was probably the content essentially of [Lev] Pontryagin's dissertation, ... he will send you the manuscript or page proofs, so that Moscow will be quoted correctly. As for topologists belonging to the guild, only H. Kneser was there, who you don't even count as a guild member![9]

Clearly, by this time Emmy Noether had her finger on the pulse of current developments in the fast-breaking field of topology.

The relationship Bartel van der Waerden forged with Emmy Noether during the 1920s (Fig. 5.1) was of decisive importance for his career. Largely due to the success of his two-volume textbook, *Moderne Algebra* [van der Waerden 1930/31], he later came to be regarded as one of Noether's leading disciples, a role he played only reluctantly. This book project actually dated from the year 1926/27 when van der Waerden was a Rockefeller fellow in Hamburg and attended Emil Artin's lecture courses. Artin had informed Courant that he and van der Waerden would write a textbook on abstract algebra based on these lectures, a plan that Artin

---

[7]On the significance of Kneser's work in this field, see [Hofmann/Betsch 1998, 9–11].

[8]In combinatorial topology, one studies triangulated spaces. It had been conjectured that if a space has two different triangulations, then these have refinements which are equivalent, but this turned out not to be true in general for dimensions four and higher; see [Scholz 2008, 865–866].

[9]Noether to Alexandrov, 13 October 1929, translated from [Tobies 2003, 103].

soon dropped after realizing that his coauthor was far more enthusiastic about this project than was he [Schneider 2011, 100–102].

B.L. van der Waerden was most definitely fascinated by Emmy Noether's mathematical vision, even if he never entirely shared her enthusiasm for abstract algebra as an end unto itself. In fact, his original interests were closer to those of Max Noether, whose work he studied in Amsterdam, and one may fairly doubt that he ever felt completely at home with the purely algebraic direction that Emmy Noether promoted.[10] Van der Waerden may have spent part of his career in Noether's shadow, but he was never her epigone; in fact, his mathematical interests were far broader than hers. In Göttingen, he also deepened his knowledge of physics by taking Courant's lecture course on methods of mathematical physics, taught from the newly published volume [Courant/Hilbert 1924].[11]

Still, it is undeniable that Emmy Noether played a formative part in van der Waerden's mathematical education, and he was the first to concede that he learned modern ideal theory, among several other things, from her. During his first year in Göttingen, he took her course on group theory and hypercomplex numbers (the older term for what later were called algebras). Already by this time, she was lecturing on Joseph Wedderburn's structure theory of algebras over arbitrary fields, thereby laying the groundwork for what would later become her principal field of research.

In a conversation with Auguste Dick, van der Waerden recalled a memorable scene from one of Emmy Noether's lectures when she wanted to present a new proof of Maschke's theorem in group representation theory. This was a classic result from the late 1890s, thus from well before the time when abstract algebra came into vogue. Not surprisingly, Noether was keen to present it from a modern point of view. In all likelihood, she had come up with the main ideas for her new proof not long before arriving in the lecture hall. She was, even more than usual, in a buoyant, upbeat mood when she began setting out her argument on the blackboard. Somewhere along the way, though, her mood changed, as it dawned on her that there was a serious hole in her proof that she was not going to be able to repair after a short moment's thought. To the surprise of her audience, she suddenly broke into a rage, threw down her chalk, and stomping on it yelled out: "Now I'm forced to do it the way I don't want to!" She then picked up a new piece of chalk and presented the traditional proof flawlessly [Dick 1970/1981, 1981: 39–40]. As this anecdote shows, Emmy Noether had a complete mastery of the methods and results of classical algebra.

During his initial stay in Göttingen, van der Waerden had already begun to grapple with problems in the foundations of algebraic geometry that would

---

[10]For a probing look at van der Waerden's views on algebraic geometry over the course of his career, see [Schappacher 2007].

[11]This interest bore fruit later when he published his monograph [van der Waerden 1932] on group-theoretic methods in quantum mechanics, a study that appeared in Courant's "yellow series" immediately after *Moderne Algebra*. For a detailed account of van der Waerden's physical interests and his mathematical contributions to quantum physics, see [Schneider 2011].

Figure 5.1: B.L. van der Waerden and Emmy Noether, Göttingen, Summer 1929 (Auguste Dick Papers, 13-1, Austrian Academy of Sciences, Vienna)

dominate his attention for many years to come [Schappacher 2007]. Already in Amsterdam, he had studied invariant theory and its importance for geometry, a viewpoint made famous by Felix Klein in his "Erlangen Program" [Klein 1872]. But as he later recalled, "when I studied the fundamental papers of Max Noether ...and the work of the great Italian geometers, ...I soon discovered that the real difficulties of algebraic geometry cannot be overcome by calculating invariants and covariants" [van der Waerden 1975, 32]. A swarm of related questions soon filled van der Waerden's mind, one of them a question Emmy Noether already addressed in her report for the German Mathematical Society [Noether 1919, 197–201], namely

> ...the generalization to n dimensions of Max Noether's "fundamental theorem on algebraic functions." ...From the papers of Max Noether I knew that this question is of considerable importance in algebraic geometry, and I succeeded in solving it in a few special cases. I did not know then that Lasker and Macaulay had obtained much more general results. [van der Waerden 1975, 32]

Van der Waerden went on to relate how Noether taught him that the tools needed to handle such questions "had already been developed by Dedekind and

Weber, by Hilbert, Lasker and Macaulay, by Steinitz, and by Emmy Noether herself" [van der Waerden 1975, 32–33]. The bookish Dutchman had no difficulty absorbing the literature she advised him to read.[12] These works opened up a whole new world for him:

> The mathematical library of Göttingen was unique. Everything one needed was there, and one could take the books from the shelves oneself! In Amsterdam and in most continental universities this was impossible. So I started learning abstract algebra and working at my main problem: the foundation of algebraic geometry. [van der Waerden 1975, 33]

Noether's paper [Noether 1923c] elaborated on a new approach to elimination theory developed by her student Kurt Hentzelt, who defended his dissertation summa cum laude in Erlangen shortly before the outbreak of the war. His official adviser, Ernst Fischer, stipulated that the text had to be rewritten prior to publication [Koreuber 2015, 15], but Hentzelt had no opportunity to do so: he was killed during the early months of the fighting.[13] The paper van der Waerden read, [Noether 1923c], was a later refinement of Hentzelt's theory. He later recalled how these ideas enabled him to give a precise notion of the generic points of an algebraic variety:

> After her lecture we, Noether's students, often discussed mathematical problems with her. My problems mainly were concerned with algebraic geometry, since I found what I had learned in Amsterdam all very nice, but knew that it lacked rigorous grounding and I was searching for such a foundation. I needed algebra for that. So I presented her with my basic problems that I had already struggled with in Amsterdam. For example: How do you define the dimension of ... an algebraic variety? One has a system of equations that define an algebraic variety in n-dimensional space. What does it mean when one calls this variety a curve or a surface? What do the Italians mean when they speak of a punto generico, a general point, of a variety? Well, ... a general point on a curve cannot be a double point or an inflection point, nor can it be the point of contact of a double tangent. In short, a general point is required to have no special properties that do not belong to all points. Is there such a thing? I found the answer to this question in a work by Emmy Noether on elimination. [van der Waerden 1997]

Soon afterward, however, van der Waerden saw that one did not need elimination theory for the concept of general points. He wrote up a paper on zero-sets

---

[12]These included Steinitz's classic paper on abstract fields [Steinitz 1910], Macaulay's Cambridge Tract on modular systems [Macaulay 1916], the famous paper of Dedekind and Weber on algebraic functions [Dedekind/Weber 1882], and Noether's own papers on ideal theory [Noether 1921b] and elimination theory [Noether 1923c].

[13]Years later, realizing the importance of Hentzelt's methods [Koreuber 2015, 98–99], Noether took it upon herself to rewrite his dissertation in [Noether 1923a], after publishing a note about this work for the German Mathematical Society [Noether 1921a].

of polynomials and then showed it to Emmy Noether, who told him it was very good and that she would submit it right away to *Mathematische Annalen*. She also gave him some tips on how to conceptualize the work and improve the presentation by changing the order of some definitions and theorems, and it soon appeared as [van der Waerden 1927]. What she never told him was that she had already come up with the same idea and had even presented it in her lecture course about a half year before he returned to Göttingen. Van der Waerden learned about this later from Heinrich Grell, who had attended that course [van der Waerden 1971, 173]. In recalling this story, a favorite that he told many times, he noted his amazement that Emmy Noether had not said a word to him about this. "She did not want to spoil the young man's joy over his discovery! Isn't that fabulous? Gauss was completely different; he spoiled the young [János] Bolyai's joy in discovering non-Euclidean geometry by writing to him: 'I've known all that for a long time'" [van der Waerden 1997].

After van der Waerden's first semester in Göttingen, Noether persuaded Courant to support his application for a fellowship from the Rockefeller Foundation, which promoted a program initiated by the International Educational Board.[14] Noether remained in close contact with van der Waerden after he returned to Holland, as we learn from a letter she wrote to Brouwer on 14 November 1925:

> You have correctly foreseen that van der Waerden would give us special pleasure! His work on algebraic manifolds (zero sets for polynomial ideals), submitted in August to the Annalen, is truly excellent, and he is now in the middle of productive work; we are corresponding the whole time with enthusiasm. Over Christmas we can continue orally; I long ago promised Alexandrov that I would visit him then and I'll be very happy to see you again soon and speak with you. (Brouwer Papers, Noord-Hollands Archief, Haarlem)

Noether's monthlong stay in the Netherlands would turn out to be both memorable and fruitful, but before taking up this thread of events we should first recount how Emmy became friends with the young Russian she named in this letter.

5.3   Pavel Alexandrov and Pavel Urysohn

Emmy Noether's first encounter with the Russian topologist Pavel Alexandrov took place during the summer of 1923. Although their mathematical interests ran along quite different tracks, which only occasionally intersected, they nevertheless shared an openness for everything new in mathematics that led to a strong

---

[14]The IEB supported young mathematicians and scientists who wished to travel abroad to pursue post-doctoral research [Siegmund-Schultze 2001]. In van der Waerden's case, the fellowship only covered nine months, as he returned to Amsterdam to finish his doctoral degree under Hendrik de Vries.

personal bond. In this broader sense, Alexandrov must be counted as one of the leading representatives of the Noether school. Both loved to "talk mathematics" in the open air, but especially at the outdoor pool in Göttingen run by the lifeguard Fritz Klic (Fig. 6.1). An indefatigable swimmer, Alexandrov later sang the praises of this particular locale in his recollections of Heinz Hopf [Alexandroff 1976]. But he also recalled the times he and Pavel Urysohn spent there when they first visited Göttingen, where they were nicknamed "the inseparables" (a play on the property of separability in topological spaces).[15] Nearly every day we could "meet Emmy Noether and Courant ... at the university swimming pool (mainly for students) on the river Leine .... There we could also often meet Hilbert but not Landau (who, when I asked him whether he bathed replied, 'Yes, every day, in my bath at home')" [Alexandrov 1979/1980, 316].

Pavel Alexandrov grew up in a refined home in Smolensk, where his father, a gynecological surgeon, was director of the local hospital. Owing to his many and varied responsibilities, the family lived in a two-story house on the grounds of the hospital. Pavel's mother was a highly educated woman, who oversaw the upbringing of her children, in particular their early training in foreign languages. Thus, Pavel and his siblings learned French and German at home. His mother spoke both fluently, but she engaged a governess from Riga, who helped instill her son's lifelong love of the German language. This woman spoke with a distinctive Baltic accent that Alexandrov later recalled when he heard Hilbert lecture.

Alexandrov studied at the renowned Moscow State University, where he came under the influence of the Soviet analysts Dmitri Egorov and Nikolai Luzin, the latter a pioneer in the field of descriptive set theory. Luzin's interests in this modern theory were first awakened in Paris, where during the Revolution of 1905 he attended lectures given by Émile Borel. He later spent three years in Göttingen as a research fellow, during which time he worked closely with Edmund Landau. After the October Revolution of 1917, Moscow State University began to admit students from proletarian and peasant families by enabling them to do preparatory work before taking their entrance examinations. At this newly reformed institution, Luzin established a famous research seminar during the 1920s whose members came to be called "Luzitania." Among them were Alexandrov and his intimate friend Pavel Urysohn, along with Aleksandr Khinchin, Andrey Kolmogorov, Mikhail Lavrentyev, Alexey Lyapunov, Lazar Lyusternik, Pyotr Novikov, Lev Schnirelmann, and many others.

Alexandrov and Urysohn soon began an intense collaboration that led to several fundamental results in general topology. Urysohn, who grew up in a Jewish family from Odessa, was an adventurous spirit. Both he and Alexandrov loved to travel, and since they both spoke excellent German, Urysohn easily convinced his friend that they should try to visit Germany in the summer of 1923. The only question, then, was how to finance their trip. Realizing that Russian intellectuals

---

[15]Felix Hausdorff invented this nickname; he addressed the two Russians as the "Herren Inséparables" in a letter from 11 August 1924 [Hausdorff 2012, 22]. They, in turn, expressed their delight with this title in a footnote to an earlier letter [Hausdorff 2012, 16].

were keenly aware of and curious about Einstein's new theory of relativity, they came up with the brilliant idea of offering public lectures on this subject at various locations around Moscow. And this venture actually paid off, or at least provided them with enough money to travel and stay abroad [Alexandrov 1979/1980, 297–298].

As budding topologists, they were particularly eager to meet Felix Hausdorff, a pioneering figure in the field. In a letter dated 18 April 1923, they announced a whole series of their new results for him, and Hausdorff reciprocated by warmly welcoming the two young Russians to visit him in Bonn. His return letter, however, dated 22 May, only arrived after they had left Moscow. Thus, they only received his message in Göttingen, a mishap due to the fact that Alexandrov and Urysohn had sent their letter to his old address in Greifswald, not realizing that he had been living in Bonn since 1921. In any event, their planned visit could not have taken place due to the Ruhr crisis, during which Bonn fell under French occupation. Despite this disappointment, they were delighted by the friendly reception that awaited them in Göttingen, a visit arranged by Landau.

In Göttingen, Alexandrov and Urysohn attended Hilbert's course on "Anschauliche Geometrie," later published as [Hilbert/Cohn-Vossen 1932]. They also heard Landau's lectures on analytic number theory and Courant's on differential equations in mathematical physics. Yet, most of all, Alexandrov remembered Emmy Noether's lectures as particularly inspiring; he also recalled her famous saying – "es steht schon alles bei Dedekind" – a view he firmly dismissed. Many decades later, he related how "[h]er lectures enthralled both Urysohn and me. In form they were not magnificent, but they conquered us by the wealth of their content. We constantly met Emmy Noether on a relaxed basis and very often talked to her about topics both in ideal theory, and in our work, which had caught her interest at once" [Alexandrov 1979/1980, 299]. Noether's lively enthusiasm for mathematical discussions left a deep impression on Alexandrov and Urysohn, both of whom were highly sociable:

> We were constantly meeting Emmy Noether on her famous walks which were first called algebraic and after our arrival came to be called topological algebraic. There were always many young mathematicians taking part in these walks, which were a model for the topology walks of our Moscow topology seminar, but of quite a different character. [Alexandrov 1979/1980, 316]

When Brouwer learned that Alexandrov and Urysohn were planning to return to Göttingen during the summer of 1924 as well as meet with Hausdorff in Bonn, he was more than willing to help make arrangements for them to visit him in the Netherlands. Brouwer's earlier work stood closer to traditional combinatorial topology, which derived from geometrical investigations of polyhedra, whereas Hausdorff's investigations grew out of Cantorian set theory and were far more general and systematic. The latter's *Grundzüge* sparked a tremendous outburst of new researches in Poland and Russia, including those of Alexandrov and Urysohn,

who helped launch the field of continuum theory.[16] Their meeting with Hausdorff was, indeed, nothing less than an historic event in the history of modern
topology, marking the beginning of Alexandrov's friendship with Felix Hausdorff,
as documented in their correspondence from 1924 up until 1935 [Hausdorff 2012,
20–133].

After spending several weeks in Göttingen, the two Russians arrived in Bonn
on July 9. Despite warnings from Hausdorff's wife, they followed their natural
instincts and took to swimming in the Rhine River, evading the barges along the
way. Their stay was prolonged by difficulties they encountered when applying for
visas to enter the Netherlands, though this meant they had more time to discuss
their work with Hausdorff, a sharp-minded listener who took an avid interest
in their future plans. The two young Russians were also warmly welcomed in
Blaricum by Brouwer, though their host had to leave only a few days later to give
a lecture in Göttingen. During this visit, they discussed a future plan for both
to return with fellowships from the International Education Board. Around this
same time, Brouwer wrote to IEB President Wickliffe Rose in New York City:

> Two young Russian mathematicians full of promise, Dr. Alexan-
> droff and Dr. Urysohn, of Moskau, desire to study topology under me.
> I already spoke of them to you this winter on our meeting in the Amstel
> Hotel at Amsterdam, and I can readily declare that the International
> Education Board would perform an action of decided scientific interest
> by enabling these very clever young scholars to come to Amsterdam
> and live there this winter. You find their request with biographic notice
> enclosed in this letter.[17]

5.4 Urysohn's Tragic Death

After their stay in the Netherlands, Alexandrov and Urysohn spent four days
in Paris, but without seeing any mathematicians there. They then left to vacation
in Batz-sur-Mer in Brittany, where they worked together during most of the day
and then went swimming in the Atlantic. They arrived around the first of August
and sent a letter to Hausdorff, picking up on their last conversations in Bonn, but
also reporting briefly on their days together with Brouwer. They were particularly
happy to have seen that, despite his present focus on foundational studies, Brouwer
remained keenly interested in other parts of mathematics, including topology in
general spaces. Looking back on all they had experienced during the past months,
the two highlights had been their visits in Bonn and Blaricum and their conversa-

---

[16]On Hausdorff's *Grundzüge*, see [Purkert 2002]; topological continua are compact connected
metric spaces.

[17]Brouwer to Rose, August 1924, Rockefeller Archive Center, IEB Collection, 1-1, Box 44,
Folder f616, Tarrytown, New York.

tions there with Hausdorff and Brouwer, respectively.[18] Indeed, in the course of two weeks they had won over two of the leading coryphaei in the field of topology.

And then, just a fortnight later, tragedy struck. On a Sunday, August 17, 1924, they were caught in a storm that took Urysohn's life, a horrible event for all who had known him. For a long time afterward, Alexandrov could not believe what he suddenly lost on that day. In the years that followed, his friendship with Brouwer was strengthened by their common interest in preserving Pavel Urysohn's legacy [van Dalen 2013, 424–434]. Some forty years later, Alexandrov described that fateful day in detail:

> The main part of the day was spent on work, and in spite of our custom it was already five o'clock in the afternoon when we got ready to go swimming. When we got into the water, a kind of uneasiness rose up within us; I not only felt it myself, but I also saw it clearly in Pavel. If only I had said, "Maybe we shouldn't swim today?" But I said nothing
> . . . .
>
> After a moment's hesitation, we plunged into a not very large shore wave and swam some distance into the open sea. However, the very next sensation that reached my consciousness was one of something indescribably huge, which suddenly grabbed me . . . . A moment later I came to myself on the shore, which was covered with small stones - it was the shore of a bay, separated from the open sea by two rocks between which we had had to swim as we made our way to open sea.
>
> I had been thrown over by a wave, right across these rocks and the bay. When I was on my feet, I looked out to sea and saw Pavel at those same rocks, already in the bay, passively rolling on the waves (which were comparatively small in the bay) in a half-sitting position. I immediately swam up to him.
>
> At that time I saw a large group of people on the shore. . . . After swimming to Pavel, I put my right arm around him above his waist, and with my left arm and my legs I began to paddle to shore with all my might. This was difficult, but no one came to my assistance. Finally, when I was already quite near the shore, someone threw me a rope, and within a few moments I reached land. Then eye-witnesses told me that the same great wave that had thrown me across the bay had struck Urysohn's head against one of the two rocks and after that he had begun to roll helplessly on the waves in the bay.
>
> When I pulled Pavel to the shore and felt the warmth of his body in my hand, I was in no doubt that he was alive.
>
> Some people then ran up to him, and began to do something to him, obviously artificial respiration. Among these people, there happened to be, as I was later told, a doctor, who apparently directed the attempts at life-saving. I do not know and did not know then how long

---

[18] Alexandrov and Urysohn to Hausdorff, 3 August 1924, [Hausdorff 2012, 20–21].

they continued, it seemed like quite a long time. In any case, after some time I asked the doctor what the condition of the victim was and what further measures he proposed undertaking. To this the doctor replied "Que voulez vous que je fasse avec un cadavre?" ...

Some more time passed, and I went into my room and finally dressed. (Until then I had remained in my swimming clothes.) Pavel Urysohn lay on his bed, covered by a sheet; there were flowers at the head of the bed. It was here that I thought for the first time about what had happened. All my experiences, all my impressions of that summer, and indeed of the last two years, rose up in my consciousness, with such distinctness and clarity. All this merged into a single awareness of how good, how exceptionally good, things had been for each of us, only about an hour ago.

And the sea raged. Its roaring, its crashing, its bubbling, seemed to fill everything. [Alexandrov 1979/1980, 318–319]

The next day, Alexandrov sent telegrams to Brouwer and to his brother, who informed the Urysohn family about what had happened. The funeral took place the day afterward, on 19 August; Alexandrov contacted a local rabbi, who performed the funeral rites. He then left Batz the next day, spent the following day in Paris, and arrived in Göttingen on the 22nd, where he was met by Brouwer, Courant and Emmy Noether. Hilbert and Klein also requested that Alexandrov visit them.

On the evening before he departed for Moscow on the 3rd of September, Alexandrov wrote an emotional farewell letter to Hausdorff from Berlin.[19] He apologized for writing so directly and personally about how he felt, but he needed to describe his state of mind and how his euphoric life had suddenly turned to misery with Urysohn's death. Beyond that, he wanted Hausdorff to know that his friend had left behind a great deal of important work; he assured him that one paper, in particular, once it appears, would truly astonish Felix Hausdorff by showing him "how deeply [Urysohn] could penetrate into the very most hidden secrets of topological space structure." This would now become Alexandrov's next great task, to prepare Urysohn's posthumous papers for publication.

5.5 Helping a Needy Friend

Upon returning to Moscow, the distraught topologist received a consoling message from Emmy Noether, in which she addressed him as "my dear, poor Paul Alexandrov" and with the familiar "Du" form. She had wanted to suggest that they use "Du" already in Göttingen, but as she explained, she was not able to summon up the courage during his short last visit. Recalling her two Russian friends, "the inseparables," she wrote:

---

[19]Alexandrov to Hausdorff, 2 September 1924, [Hausdorff 2012, 27–28].

I have the image of both of you always in mind, with all that outpouring of life that you brought forth; and now you are all alone. But he lives on with you; and if his manuscripts now bring his thoughts back to life – and only you can bring them to full life – so he will return to you more and more. And the sharp pain will perhaps not hurt so much after all; and you can always then think back with more gratitude for what you had during these four years.[20]

During the next three years, Alexandrov attached himself closely to Brouwer, who also took a deep interest in preserving Pavel Urysohn's legacy. Brouwer arranged an invitation for Alexandrov to Amsterdam with financing through the Rockefeller Foundation and the IEB. One month after Urysohn's death, he wrote to Rose:

I have to inform you that Doctor Urysohn, one of the two Russian mathematicians I proposed to you for a fellowship of the International Education Board, found a sudden death by a most tragical accident some weeks ago. He left a mass of posthumous scientific papers and notes which when elaborated and edited will prove, I am certain, to contain results of the highest scientific importance. I believe that this elaboration will come to the best end, if it is undertaken by Urysohn's friend Doctor Alexandroff under my guidance. This is also the opinion of Doctor Alexandroff, whose desire to come to me to Amsterdam is in this way strengthened by a motive of scientific piety.[21]

Arriving in May 1925, the young Russian took up residence in the village of Blaricum, where Brouwer lived and spent most of his time. Blaricum long attracted an assortment of artists and elites, many of whom worked in nearby Amsterdam. Brouwer owned a good-sized house there, but he spent most of his time in a small cottage with a desk and piano, situated in a fairly large, but completely overgrown garden. This is where photos of Brouwer and his lively Russian friends were taken the year before (see Fig. 5.2), and where Alexandrov and Brouwer began their work editing Urysohn's posthumous papers.

Later that year, Alexandrov received an inquiry from Noether regarding plans for his forthcoming stay in Göttingen. She was eager to have him offer a course on topology during the summer term, and he was keen to reciprocate, as he was about to begin lecturing on the subject in Amsterdam. He replied to her from Blaricum as follows:

...I hope not only to give a course on topology [in Göttingen] but also to regularly follow your lectures on the foundations of group theory. You know that I am very interested in your works in this area, especially because in terms of their content and methodology they are very

---

[20]Noether to Alexandrov, 1 September 1924, transcribed in [Tobies 2003, 102].

[21]Brouwer to Rose, 17 September 1924, Rockefeller Archive Center, IEB Collection, 1-1, Box 44, Folder f616, Tarrytown, New York.

Figure 5.2: Pavel Alexandrov and Pavel Urysohn, Blaricum 1924 (Hausdorff Papers, Bonn University)

close to the circle of ideas in general topology. I am also counting on stimulation from your side because, as you also know, the many mathematical discussions that Urysohn and I had with you were among the most lively and stimulating we have ever had, as I've told Brouwer on several occasions. Brouwer also knows how excited I was about your correspondence from last summer about group theory. So I ask you to count on me to be an attentive and eager auditor.[22]

Noether had discussed with Courant whether it might be possible to extend Alexandrov's IEB Fellowship, a plan that turned out not to be feasible. In any event, Alexandrov cautioned her that she should take this up with Brouwer, whose prickly and oftentimes domineering personality made the Russian very wary:

I consider it necessary to write you that I feel obliged to let this whole question rest with Brouwer's decision: for moral reasons it would be quite impossible for me to accept something in this direction if Brouwer

---

[22]P.S. Alexandrov to E. Noether, 11 November 1925, Hochschularchiv der ETH Zürich, Hs 160.

does not fully agree. ...I don't, in fact, want my purely moral dependence on Brouwer regarding these and some other questions, which I take on freely, and which consequently in no way violates my human and scientific freedom, to lead to any misunderstanding. ...I don't want to influence Brouwer's decisions in any way, because I want to avoid at any price the possibility of internal friction with Brouwer.

Three days later, Noether posted the letter already cited above, in which she sang the praises of van der Waerden. Along with it, she enclosed Alexandrov's letter, and her own wishes for his visit:

> I would like to write you today about Alexandrov's planned stay in Göttingen during the coming summer and the possibility of extending his Rockefeller scholarship. He wrote me about his plans and views in the attached letter, which he authorized me to show you.
>
> I would be so pleased if Alexandrov received an extension of his scholarship as much as I wish him anything at all that would make his life a little easier! I don't think any formal reason stands in the way either, since he wishes both to learn as well as teach here just as in Amsterdam. Would it perhaps be possible to proceed as with van der Waerden: that you present the proposal and I countersign it, since Alexandrov wants to work with me? Of course, any other arrangement that you might propose would be fine with me. It would also be possible to submit an application from here; Trowbridge[23] is said to have been generally very accommodating during a visit a few weeks ago, though I didn't speak with him myself. In any case, I hope that either way the matter comes to a good end, and may I ask you to inform me of your intentions soon?[24]

As Noether soon learned, Brouwer was more than happy to cooperate with her in support of this plan. She was undoubtedly delighted to read his response: "You will surely experience a great joy in Göttingen from Alexandrov's lectures; he recently began a course of lectures here that has completely captivated his auditors, not only due to the clarity and precision of the subject matter but also because of his enthusiasm as a lecturer."[25] Brouwer went on to say that he would have been happy to promote Alexandrov's career chances in the Netherlands were the latter not already so advanced that he could expect to gain a professorship in Russia.

So these plans for Alexandrov's stay in Göttingen had already been settled by mid-December when Emmy Noether came to spend nearly a month in Blaricum.

---

[23] Augustus Trowbridge was an experimental physicist from Princeton, who was appointed as the IEB's chief representative in Europe; his first visit to Göttingen in October 1925, mentioned by Noether here, is described in [Siegmund-Schultze 2001, 145–148].

[24] E. Noether to L.E.J. Brouwer, 14 November 1925, Brouwer Papers, Noord-Hollands Archief, Haarlem.

[25] Brouwer to Noether, 21 November 1925, Brouwer Papers, Noord-Hollands Archief, Haarlem.

She stayed until January 10, residing in the village's Villa Cornelia.[26] During her stay in Blaricum, Noether connected again with Alexandrov and van der Waerden, while meeting other mathematicians in Brouwer's circle. Alexandrov offered a vivid recollection of a dinner party in her honor at Brouwer's home, "during which she explained the definition of the Betti groups of complexes, which spread around quickly and completely transformed the whole of topology. When Emmy Noether arrived at Blaricum, her student van der Waerden, who was then 22 years old, also came. I remember extraordinarily lively mathematical conversations in which he took part" [Alexandrov 1979/1980, 324].

Emmy Noether never published anything directly related to algebraic topology, and yet she is rightly regarded as a central figure in the early history of this field. Already in January 1925, she spoke about homology groups in in Göttingen. This talk took place a full year before Emmy Noether's visit in Blaricum, so there is every reason to treat Alexandrov's anecdote as an accurate account. Noether clearly exerted a strong influence on Alexandrov and Hopf, who would soon become two of the leading figures in the field. Already in a paper Alexandrov submitted to *Mathematische Annalen* in November 1925, he added the footnote: "The first, abstract part of the present work is closely related to the more recent studies by Miss E. Noether from the field of general group theory and is partly inspired by these studies" [Alexandroff 1927, 555].

During his stay in Blaricum, Alexandrov was able to finish most of the editorial work on Urysohn's posthumous papers. Afterward, from the beginning of May until mid-July 1926, he was in Göttingen residing as Emmy Noether's house guest at Friedländerweg 57 (Fig. 6.3). From there, Alexandrov wrote to Hausdorff on 13 May 1926, informing him that he had received the first set of page proofs for the new edition of Hausdorff's book on set theory.[27] Alexandrov promised to return the proofs with his comments in the next two or three days, adding that these elements of set theory had by now crystallized to the point that there was hardly room for any substantial *mathematical* differences of opinion.

I thank you again very much for giving me the opportunity to be one of your first readers. By the way, I notice during my present lecture course in Göttingen that I can already cite your first edition by heart (just like good conductors, for example, who conduct the Beethoven symphonies without the score!).

I'm very happy to be teaching a systematic course on topology here .... The interest in the whole complex of questions in set-theoretic topology is *very* great, above all for Urysohn's theory of dimension; Urysohn himself only experienced the initial stage of this interest in

---

[26]These details surface in a letter she wrote to Einstein on January 7, in which she advised him about a paper that had been submitted for publication in *Mathematische Annalen*.

[27]*Mengenlehre*, published in 1927, reprinted in [Hausdorff 2008, 1–351]. This was actually an entirely new book, as described in [Hausdorff 2002, Hausdorff 2008].

and recognition for his truly most significant mathematical achievement. [Hausdorff 2012, 42]

Toward the end of the semester, Noether and Alexandrov organized a small meeting on group theory and topology, or as Alexandrov called it in a postcard sent to Hausdorff, a "Locarno conference" that had reached a peaceful conclusion at a favorite watering hole just outside Göttingen, the "Kehr" restaurant on Hainholzweg.[28] Some weeks earlier, Alexandrov sent Hausdorff his reaction on reading the proofs of his new book *Mengenlehre*. Unlike the *Grundzüge der Mengenlehre*, Hausdorff's new book dealt almost exclusively with metric spaces, a restriction Alexandrov apparently found disappointing. He expressed his general òpinion in a letter from July 4, in which he also sketched his own research program for the future, namely to build bridges from point set topology à la Hausdorff to the more classical geometric topology dating back to the nineteenth century.[29] In August, he returned to Batz, where he completed the manuscript for [Alexandroff 1928], the paper in which he proved his fundamental theorem on the approximation of compact metric spaces by polytopes.[30] He described his more general goals to Hausdorff in these words:

> I'm working on various, in general really difficult questions (with still limited success!), all of which concern filling the long-standing deep gorge separating general (set-theoretic) and classical topology, and hoping thereby to be able to clarify many topological features of our ancient, God-given physical space.
>
> I'm grateful for a great deal of stimulation in pursuing all these plans – for which some parts require entirely new methods – to the young Berlin mathematician Hopf, a truly outstanding topologist who was here during this summer semester (he has carried some of Brouwer's things further in remarkable ways and has completed them with essentially new methods); he is also personally a very nice human being, which is of course a necessary condition for purely scientific relations. [Hausdorff 2012, 44]

Alexandrov somehow neglected to mention one other (for him) important attribute of Heinz Hopf: he, too, was an avid swimmer.

---

[28][Hausdorff 2012, 45]; among the others who signed this postcard were Brouwer, Landau, Grell, and Erich Bessel-Hagen.

[29]In a broad sense, this gulf was never completely bridged; see [Brieskorn/Scholz 2002].

[30]The bridge Alexandrov built in [Alexandroff 1928] was based on his notion of nerves of coverings, a novel way to pass from coverings of a compact subset of a metric space to polytopes, the traditional objects studied in combinatorial topology; see [Hurewicz/Wallman 1948, 67–72].

Chapter 6

Noether's International School in Modern Algebra

6.1   Mathematics at "The Klie"

Pavel Alexandrov and Heinz Hopf met for the first time in Göttingen in the spring of 1926, soon after Alexandrov departed from Blaricum. Hopf had recently taken his doctorate in Berlin under Ludwig Bieberbach and Erhard Schmidt, and his research interests differed sharply from Alexandrov's work in general topology. As Hopf and Alexandrov gradually discovered their common interests over the course of that summer, they soon became close friends as part of Emmy Noether's circle. She knew, of course, that no one could ever replace Pavel Urysohn in Alexandrov's life, but she surely felt that this new friendship with Hopf was a turning point for him. Noether's disarming frankness and warmth spilled over very quickly whenever outsiders came to Göttingen, creating a memorable atmosphere for those who shared her addiction for mathematics and the simple pleasures of life. Such as swimming. Hilbert had long been an inveterate swimmer, going back to his years in Königsberg when he often spent vacations at nearby locales on the Baltic. His legendary quip in rebuking those who opposed Noether's habilitation – "we're a university, not a bathing establishment" – takes on more vivid meaning in the light of how important swimming was for many of the mathematicians in Göttingen.[1] And when they thought about topology while swimming, then very likely Pavel Alexandrov and Heinz Hopf were in their midst.

B.L. van der Waerden, who had picked up some of the latest trends in topology two years earlier on afternoon walks with Hellmuth Kneser, now eagerly listened to the latest ideas that would soon take form as standard concepts in algebraic topology. Decades later he recalled one of these conversations, in which Hopf, Alexandrov, and Emmy Noether discussed Lefschetz's fixed point formula.

---

[1] This was pointed out by Cordula Tollmien in her delightful essay [Tollmien 2016b].

D. E. Rowe, M. Koreuber, *Proving It Her Way*, https://doi.org/10.1007/978-3-030-62811-6_6

Figure 6.1: "The Klie": a breeding locale for mathematics and mosquitos (Städtisches Museum Göttingen)

This formula, which had just been published by Lefschetz, made it possible to calculate the number of fixed points of a continuous map, or more precisely, the sum of the indices of the fixed points for a mapping of a manifold into itself. Emmy Noether said that one should not do this with matrices, not by calculation, but with concepts, with additive groups and homomorphisms of these groups. Then everything becomes much more transparent and beautiful. And so the old concepts, such as Betti numbers and torsion numbers, were to be retained, but based on group theory. The basic concept from which these older ones were derived was that of a homology group, a notion familiar to every topologist today. ...When Lefschetz's fixed point formula was later formulated and proven by group theory, Emmy Noether was thrilled. In such cases Emmy Noether ... sometimes used to say: "The proof is now conceived abstractly and thereby made transparent." For her, that was the point of modern, abstract algebra: to avoid all special calculations, matrices, etc. and through abstraction to do away with all insignificant features of a particular problem in order to make the essential concepts visible by placing these up front so that the entire proof becomes transparent. [van der Waerden 1997]

Years later, in his personal memories of Heinz Hopf, Pavel Alexandrov recalled the lively atmosphere that both of them enjoyed that summer, including musical evenings with the Courants, boat rides on the Leine River, and afternoons spent swimming and chatting at "the Klie" (Fig. 6.1), this being the colloquial name for the university swimming pool run by its lifeguard, Fritz Klie, located next to the river and just south of the town.

Many a mathematical, but not only mathematical conversation took place at the Klie, either in the moving waters of the Leine, which were not always particularly clean, even quite brown after it had rained, or in the sun or else the shade of the lovely trees, a favorite spot for the mosquitos. And many a mathematical idea was born there as well. ... The Klie swimming pool was exclusively for men; females were only represented by Miss Emmy Noether and Mrs. Nina Courant, both of whom exercised their exclusive privileges on a daily basis, no matter what the weather conditions. [Alexandroff 1976, 114]

Van der Waerden was also a regular guest at the Klie. In a letter to Felix Hausdorff, Alexandrov reported on an interesting new paper by David van Dantzig and van der Waerden, which also happened to betray the influence of Göttingen's summer culture: "for the first time in the mathematical literature a certain surface (a sphere with three holes) has been 'officially' designated *the swimsuit*. The stimulus for introducing this terminology can no doubt be traced to activities connected with my two-semester topology seminar in Göttingen."[2]

Hilbert, Courant, Otto Neugebauer, and numerous others were also often to be seen at the Klie, and they were joined that summer by Brouwer as well. Two years earlier, in the midst of the *Grundlagenstreit* with Hilbert, Brouwer had been invited by the philosopher Moritz Geiger to give a lecture on intuitionism in Göttingen. He spoke at that time before a large crowd, few of whom sympathized with his position. According to Hans Lewy, who heard this lecture as a doctoral student, a heated discussion followed. Hilbert did not take part until the very end. Then he stood up and said: "With your methods most of the results of modern mathematics would have to be abandoned, and to me the important thing is not to get fewer results but to get more results" [Reid 1970, 184]. The audience's response to this retort was predictable – thunderous applause – after all, this was Hilbert's crowd. Still, this could hardly pass as a defense of formalism, which Brouwer had long been attacking. Hilbert was merely expressing a pragmatic view, one that appealed to many mathematicians, like Hans Lewy, who simply wanted to carry on with their research projects.

---

[2] Alexandrov to Hausdorff, 20 December 1928, [Hausdorff 2012, 78].

A similar sympathy for Hilbert's position can be seen from letters Alexandrov wrote to Hausdorff during this time.[3] Four months later, Alexandrov assured Hausdorff (who detested intuitionism[4]) that he remained a formalist and fully accepted Georg Cantor's credo that "the essence of mathematics lies in its freedom." Even though he did not share Brouwer's views, he still had high respect for his position, which he regarded as an heroic attempt to draw earthly knowledge from mathematics, whereas he saw the latter as primarily an art form. From a neighboring hotel room in Berlin, Alexandrov was listening to the funeral march from Beethoven's "Eroica" symphony, which led him to express the thought that it contained more real knowledge than all of science and mathematics, formalism and intuitionism included.[5]

During Brouwer's visit in 1926, Emmy Noether and Alexandrov were eager to restore harmonious relations between Hilbert and the Dutch topologist turned philosopher, who evidently felt warmly received by those closest to Courant and Noether. She decided to invite a group of mathematicians, including Hilbert and Brouwer, to her apartment one evening (Fig. 6.3). Those present included Courant, Landau, and Hopf, all seated around Emmy's table in her cozy quarters. She and her Russian friend decided that he should think of a good way to break the ice between Hilbert and Brouwer, and Alexandrov decided to try an age-old expedient: what better way to bring two persons together than by innocently mentioning the name of a third person who was *persona non grata* for both? The list of potential candidates might well have been long, but none could compete with the widely disliked and famously vain Leipzig mathematician Paul Koebe, who became notorious in Göttingen for stealing other people's ideas – a practice many there normally viewed as rather harmless. (In fact, appropriating ideas that found their way into the mathematical atmosphere in Göttingen became so commonplace that a special term was adopted for this: this was called "nostrifying" the thoughts of another.) Brouwer had tangled with Koebe already in 1911 at the Karlsruhe conference, where he first met Emmy Noether.[6] Alexandrov could hardly believe his luck once he had merely dropped this fellow's name. Before long, Brouwer and Hilbert were trying to outbid each other with nasty remarks about Koebe's character, all the while nodding agreement and warming to their heartfelt consensus, which finally ended in a toast to everyone's health for the betterment of mathematics [Alexandroff 1976, 115]. After their stay in Göttingen,

---

[3] Alexandrov expressed dismay when he learned that his former mentor Luzin took the "ignorabimus" position regarding certain difficult mathematical questions, a view Hilbert had famously rejected when he spoke about mathematical problems at the 1900 ICM in Paris (Alexandrov to Hausdorff, 29 November 1925, [Hausdorff 2012, 38]).

[4] In a letter to Abraham Fraenkel, Hausdorff wrote: "Both you and Hilbert treat intuitionism with too much respect; one must for once roll out heavier guns against the senseless destructive anger of these mathematical Bolshevists!" [Hausdorff 2012, 293].

[5] Alexandrov to Hausdorff, 4 April 1926 [Hausdorff 2012, 38–40].

[6] Brouwer's ensuing difficulties with Koebe were legendary; for details see [van Dalen 2013, 175–192] and [Rowe 2018b, 266–290].

Brouwer joined Alexandrov on his nearly annual pilgrimage to Batz, where they commemorated the genial and much-missed Pavel Urysohn.

6.2 The Takagi Connection

In 1927 two young Japanese mathematicians arrived in Göttingen to study under Emmy Noether. These were Zyoiti Suetuna and Kenjiro Shoda. Both had studied number theory under Teiji Takagi in Tokyo. Suetuna's principal interests centered on analytic number theory, so he worked closely with Landau, but also later with Artin in Hamburg as he was deeply attracted to his general reciprocity law. When he returned to Tokyo University, Suetuna introduced a research seminar modeled on those in Göttingen, and in 1936 he was appointed to Takagi's chair after the latter's retirement.

Shoda had already become interested in algebra as a student, so he chose to spend the year 1926/27 working with Issai Schur in Berlin.[7] He had never heard of Emmy Noether until the summer of 1927, so his plan to leave Berlin and travel to Göttingen had nothing to do with her. Shoda wanted to study at this famous university, but especially to hear Hilbert's lectures, knowing that he had been Takagi's teacher. When he arrived, he went to look for Emmy Noether's apartment in the Friedländerweg. Once he found it and rang, a badly dressed stocky woman came to the door and so he asked if Prof. Noether was at home. "I'm Noether and you must be Herr Shoda," came the reply, quite to his amazement. Then, as soon as he sat down, she bombarded him with questions in a very friendly way. He began to realize that Schur must have written her, because she already seemed to know various things about him. She advised him to read works by Steinitz and Krull as background for her lectures; it was very good, she told him, that he had studied representation theory in the style of Frobenius and Schur. He would soon learn about her approach to the subject.

Kenjiro Shoda found Noether's lecture style very difficult to follow, but the study tips she had given him proved extremely useful to him, especially those concerning abelian groups. Learning this material provided him with concrete examples that enabled him gradually to grasp the general concepts, which Emmy Noether only rarely illustrated by means of simple cases. In 1975, Shoda recalled one of Noether's famous sayings concerning methods of proof in mathematics. She maintained that one could, of course, prove the equality of two real numbers a and b by first showing that $a \geq b$ and then that $a \leq b$, but one should not be satisfied with such a proof. Instead, one should search for the true reason that actually reveals why a and b must be equal. One must imagine that she repeated this cryptic advice nearly as often as her most famous saying, namely

---

[7]The following information is taken from Shoda's personal recollections, which he wrote for the Japanese edition of Auguste's Dick's biography of Emmy Noether [Dick 1970/1981], published in 1975. Shoda's text was translated into German and can be found in Dick's literary estate in the Austrian Academy of Sciences, item 10-26.

"that's already in Dedekind" (das steht schon bei Dedekind).[8] Yet Shoda came to understand that these favorite remarks had a much deeper meaning, namely, that for Emmy Noether understanding a mathematical result only began with a provisional argument or the sketch of a proof. Her lectures were often confusing and difficult to follow – not only for foreigners trying to read her handwriting (she wrote on the blackboard using the old Gothic script), but for anyone, because she was often still struggling to clarify the arguments for herself.

During his year in Göttingen, Shoda rented a room in a boarding house just down the street from Noether's apartment. In fact, his landlady was on friendly terms with his teacher, who would occasionally drop by. Best of all, he often had the chance to accompany Emmy Noether on the fairly long walk from her residence to the *Auditorienhaus* where she taught. On such occasions, she normally did all the talking, and what she had on her mind, of course, was the lecture she was about to deliver. Listening to her ramble on without any visual aids at all was definitely a challenge, but it could also serve to sharpen a person's ability to think about abstract mathematics. After he grew accustomed to her speaking style, Shoda found that such "preview" performances – which were almost like hearing her give the lecture twice – contributed greatly to his understanding. Repeat performances later became standard fare for Emmy Noether in the United States, where she would lecture on Monday at Bryn Mawr College as a warm-up for the teaching she would do the next day in Princeton.

Shoda was one of only around ten auditors who attended Noether's course. Others included his compatriot Suetuna, Levitzki, and van der Waerden, who often interrupted to ask questions as he was serving as Noether's official notetaker. The material she taught that semester on "Hypercomplex numbers and representation theory" was by no means new; van der Waerden had attended the same course three years earlier. This time, though, she was ready to put everything in better shape, and so she enlisted his help in writing the lectures up for publication. Shoda and many others would later have the opportunity to study this theory in print form in [Noether 1929a].[9]

Like most of those who fell under Noether's wing, Kenjiro Shoda had to learn a new way to think about mathematics. He already knew how to read and study, and he quickly learned how to listen and "talk mathematics," but he also gained some important insights from her about writing, advice he clearly took to heart. One such insight she imparted was actually a remark she had heard from Hilbert, and since he loved to package everything in threes, we can easily imagine that this bit of wisdom originated with him. According to Hilbert, mathematicians should realize that there are three types of papers: those that no one reads, others that people skim through quickly, and then there are the rare few that other mathematicians actually study carefully. This being the case, an author need only be concerned about papers in the second category, those that will be browsed

---

[8] One finds the exact same admonition in [Weyl 1968, 3: 442].

[9] Some of its most novel features were described by Nathan Jacobson in [Noether 1983, 17–18].

rather than read. For such publications, authors should make sure to have a clearly written introduction that informs the reader about what the remainder of the paper contains (the part no one is likely to read).

Noether also informed Shoda that Schur had criticized a manuscript he had written, a remark that suggests she thought Shoda would not be offended if she imparted constructive criticism. One piece of common-sense advice he remembered all his life. She told him, "an author should not try to include all of his or her results in a single paper. Mathematical research is a lifelong activity and the publications are only signposts along the way. One should leave some things out that one can continue to work on later." Through this type of advice, he reflected, Noether gave him the foundation he needed to become a research mathematician.

They remained in contact from a distance after 1929, when Shoda returned to Japan. Soon afterward, he began to write his textbook *Abstract Algebra*, which was first published in 1932 and then reprinted many times later. Its impact on mathematics in Japan has sometimes been compared with that of van der Waerden's *Moderne Algebra* on Europe and the United States [Sasaki 2002, 246]. In 1933 Shoda was appointed as professor in the Faculty of Science at the newly founded Osaka University. After the Second World War, he was elected to chair the Mathematical Society of Japan, and in 1949 he became Dean of the Faculty of Science. He served in that office for six years, after which he was appointed President of Osaka University, a position he held for another six years. He is remembered today by students and alumni there as the founder of the Shoda Cup, awarded each year to the winning team in an athletic competition.

6.3 Bologna ICM and Semester in Moscow

It was during the academic year 1927/28 that Alexandrov's friendship with Hopf really intensified. Both were then Rockefeller fellows in Princeton, by now the leading international center for topology, led by Lefschetz, Veblen, and J.W. Alexander. During this time, Alexandrov and Hopf began planning a multi-volume work on topology, though only the first volume was completed and did not appear until 1935. They spent the following summer in Göttingen, much of the time together with Otto Neugebauer, who also enjoyed swimming in scenic surroundings. Toward the end of their stay, all three traveled to Bologna to attend the International Congress, the first postwar congress open to mathematicians from countries who were on the losing side of the First World War. Emmy Noether, who had attended the 1908 ICM in Rome, was not about to miss out on this event, even though many in Germany refused to attend for political reasons. Over the summer, she and van der Waerden polished up the manuscript from her winter-semester lectures, which Noether then submitted to *Mathematische Zeitschrift*, where it was published as [Noether 1929a]. This new work opened the way to the third main period in her research career.[10] Noether no doubt took great pleasure in presenting some of her first main results in Bologna [Noether 1928].

As noted, the Bologna ICM was a politically contentious event that exacerbated tensions between nationalists and internationalists within the German mathematical community. Hilbert clashed openly with Ludwig Bieberbach, one of several Berlin mathematicians who chose to boycott the congress, though Hilbert was convinced that behind the scenes Brouwer was the true ringleader behind this boycott effort, which he was determined to foil [Siegmund-Schultze 2016]. For some time, Hilbert had been fighting to stay alive after learning that he was suffering from pernicious anemia, which was then a lethal disease. His gloomy private thoughts fixated on Brouwer, whom he saw as a dangerous influence on the German mathematical community:

> In Germany there has arisen a form of political blackmailing of the worst kind: you are not a German, unworthy of German birth, if you do not talk and act as I prescribe. It is very easy to get rid of these blackmailers. You have only to ask them how long they were in German trenches. Unfortunately, German mathematicians have fallen victim to this blackmailing, Bieberbach for example. Brouwer ... has cultivated the instigation of discord among the Germans, all the more in order to set himself up as the master of German mathematics. With complete success. He will not succeed a second time.[11]

After the ICM in Bologna, Alexandrov Hopf, and Neugebauer vacationed in Italy on the shores of the Mediterranean. When they departed, Alexandrov traveled to Venice to meet with Emmy Noether, in all likelihood to discuss plans for her forthcoming stay in Moscow. She arrived there on 13 October 1928 and took up quarters in a dormitory near the Krymskii Bridge within walking distance of Moscow University. There she taught a course on abstract algebra, a novelty in the Soviet Union, as well as a seminar on algebraic geometry at the Russian Academy. One of those whom she met during this time was 20-year-old Lev Pontryagin, the blind topologist who would soon emerge as one of the giants in the field. Alexandrov asserted that he was strongly influenced by Noether's mathematics [Alexandroff 1935, 9], though most likely this influence was mediated by Alexandrov himself; no one made a stronger impression on Pontryagin than his teacher. The latter did, however, record these impressions from Noether's Moscow lectures:

> Around the beginning of my fourth year P.S. Alexandrov returned from abroad and brought Professor Noether with him. So I returned to topol-

---

[10]In the winter semester of 1929/30, she offered a course on "Algebra of hypercomplex numbers" that was written up by Max Deuring; these lectures, however, were only published posthumously in [Noether 1983, 711–762].

[11]Translated from [Rowe/Felsch 2019, 260–261]; a full account of the Brouwer-Hilbert conflict appears there in Chapters 7 and 8. Hilbert's allusion to Brouwer's earlier success refers to a propaganda campaign directed against French mathematicians, particularly Paul Painlevé.

ogy in my fourth year and also attended the lectures of Miss Noether on contemporary, modern algebra. These lectures had an impressive comprehensive quality quite different from Alexandrov's lectures, but they were by no means dry and I found them very interesting. Miss Noether spoke in German, but her lectures were understandable due to the unusual clarity of the presentation. The opening lecture of this famous German female mathematician was attended by a huge crowd.[12]

During that winter, Alexandrov was commuting back and forth between Moscow and his native Smolensk, where he taught at the Pedagogical Institute. He arranged this so that he could attend Noether's course in Moscow, making use of their conversations while he was teaching algebra in Smolensk. He described this as "a long algebra course in which apart from the obligatory material I presented the fundamentals of modern algebra (the theory of groups, rings and fields). I brought all these new ideas from Emmy Noether" [Alexandrov 1979/1980, 325]. This experiment apparently bore real fruit, since one of those who attended Alexandrov's course was A.G. Kurosh, later to become one of the leading algebraists in the Soviet Union.

The summer that followed was the first since 1923 during which Alexandrov decided not to visit Göttingen. Instead, he and Kolmogorov undertook a long journey (1300 km.) by rowboat on the Volga River. Returning on board a steamship, Alexandrov wrote a letter to Felix Hausdorff describing his trip, but also events of the recent past.

> The winter flew by quickly with much pedagogical (only a little scientific) work. You probably know that Frl. Noether adorned Moscow with her presence throughout the entire winter .... Her presence was a great joy for all of us and she performed a real service for Russian mathematics: she succeeded in creating great interest in a field that has long constituted a gaping hole in the Russian mathematical tradition, I mean algebra.[13]

During the summer semester after her return, Noether spoke about her experiences in Moscow in a "travel report" for the members of the Göttingen Mathematical Society.[14] She corresponded only sporadically with Alexandrov, though after her return from Moscow she tried to keep him informed about mathematical life in Göttingen. On the first day of vacation after the semester, she wrote with some news:

> ... van der Waerden has made many things more transparent and we now know what it is about; but we still don't know how it should

---

[12]Translated from the German in [Koreuber 2015, 172]; the original Russian was published in 1988.

[13]Alexandrov to Hausdorff, 10 July 1929, [Hausdorff 2012, 85–86].

[14]Delivered 4 June 1929, *Jahresbericht der Deutschen Mathematiker-Vereinigung* 38: 142; unfortunately no further record of what she reported seems to have survived.

work – at least in the algebraic-analytical part, a completely radical "transparency" will be necessary! And some of the "banalysis" – I don't know who invented that word – will still have to be thrown out! You'll see for yourself.[15]

How is the book progressing? At Landau pace, 40 pages per day?[16] Incidentally, Kerekjarto[17] also reappeared, in transit to Hamburg, where he performed. But Courant did not let him come up with any wishes.

Cohn-Vossen is editing Hilbert's "intuitive geometry" to make a yellow book out of it; but he has the right feeling, that he should only give notes and references to existing proofs, and otherwise add nothing. So the type of lectures you always talked about with so much joy will still remain the same.[18]

The final product, [Hilbert/Cohn-Vossen 1932], appeared with 230 figures that guide the reader through a vast range of geometrical knowledge, all of it presented informally. This was the furthest thing from Noether's mathematics, and yet she wrote about it respectfully.

6.4   Helmut Hasse and the Marburg Connection

During her third mathematical period, beginning in 1927, Emmy Noether worked closely with Helmut Hasse, a mathematician who made fundamental contributions to algebraic number theory and especially class field theory. Their paths no doubt crossed in Göttingen during 1919 when Hasse began his studies there, drawn especially by Erich Hecke's lectures on number theory. When Hecke soon thereafter accepted a chair at the newly founded University of Hamburg, however, Hasse decided to continue his studies under Kurt Hensel in Marburg. Not many serious young mathematicians would have considered leaving Göttingen to study at this much smaller university, but Hasse had picked up a copy of Hensel's *Zahlentheorie* [Hensel 1913] at a local bookstore and became enthralled with the author's treatment of p-adic numbers.

Hensel had studied in Berlin under the distinguished algebraist Leopold Kronecker, whose collected works he later edited. At the time Hasse met him, he had long been chief editor of the *Journal für die reine und angewandte Mathematik*, the Berlin journal commonly called simply *Crelle* after its founder. In short, Kurt Hensel, whose grandparents were the painter Wilhelm Hensel and the composer

---

[15]This passage is almost surely a reference to the topic of Noether's Prague lecture [Noether 1929c], discussed below.

[16]Edmund Landau's lectures were famously well organized and presented in rapid-fire fashion. He apparently composed his books at the same pace.

[17]Béla Kerékjártó was a Hungarian topologist who often visited Göttingen.

[18]Noether to Alexandrov, 1 August 1929, translated from [Tobies 2003, 103].

Fanny Mendelssohn, was a leading representative of the Berlin algebraic tradition.[19] Hasse thus entered a world quite distinct from the intellectual atmosphere familiar to Emmy Noether from Erlangen and Göttingen. Not that this proved to be a barrier; on the contrary, when they met again in the mid-1920s it took little time before their mutual interests set off fresh sparks.

Hasse revered his mentor, Kurt Hensel, under whom he completed his doctoral dissertation in 1921 and his habilitation thesis one year later. In 1925 he was already a full professor in Halle, where he assumed the chair long held by Georg Cantor. While there he became co-editor of *Crelle* alongside Hensel, whom he succeeded in Marburg when the latter retired in 1930. Hasse remained editor-in-chief of *Crelle* until the end of his life. His subsequent career, beginning with his appointment in Göttingen in 1934 as Hermann Weyl's successor, will occupy us in the final two chapters.

Kurt Hensel's fame rests on his invention of the p-adic numbers, which he had already introduced in 1897. American mathematicians could get a crash course in this new theory by reading Leonard E. Dickson's review of [Hensel 1908] in the *Bulletin of the AMS* [Dickson 1910]. In that same year, Ernst Steinitz published his fundamental paper on the the modern concept of fields, "Algebraische Theorie der Körper" [Steinitz 1910], noting how the invention of p-adic numbers influenced this work. "I was particularly stimulated," he wrote, "to undertake these general investigations by Hensel's theory of algebraic numbers (Leipzig, 1908), in which the field of p-adic numbers forms the starting point, a field that is neither a function- nor a number field in the usual sense of the word."

Another mathematician who took inspiration from Hensel's p-adic numbers was the Hungarian József Kürschák, who is credited with inventing the theory of valuations. Kürschák introduced the p-adic number field $\mathbb{Q}_p$ as the completion of the rational number field $\mathbb{Q}$ with respect to its p-adic valuation, thereby dispelling the mystery surrounding the status of Hensel's contruct. During the First World War, when he was studying under Hensel in Marburg, Alexander Ostrowski studied Kürschák's work and in 1916 he proved two fundamental theorems in valuation theory.[20] As a consequence, he showed that the p-adic number fields $\mathbb{Q}_p$ were the only possible completions of the rational numbers besides the usual completion leading to the real numbers $\mathbb{R}$. Ostrowski sent a preprint of his paper to Emmy Noether, who reacted with typical enthusiasm:

> I have started to read your functional equations [paper] and I am very interested in it. Is it perhaps possible to characterize the most general field which is isomorphic to a subfield of the field of all real numbers? [Roquette 2003, 6]

As Roquette noted, she was asking a question that was both natural and deep, namely which fields can be isomorphically embedded into $\mathbb{R}$.

---

[19]For insights into the long-term influence of this Berlin tradition, see [Hawkins 2000, chaps. 4, 5, 10] and [Hawkins 2013].

[20]His paper [Ostrowski 1918] was already submitted in 1916.

Whereas Noether clearly sensed the importance of these new ideas, recognition of valuation theory and the importance of p-adic number theory emerged only gradually in the wake of Hasse's work. An early turning point came in October 1920, when Hasse discovered his Local-Global Principle. This breakthrough awoke real interest in Hensel's theory, transforming it from what many in Göttingen regarded as a mere curiosity into a central tool for research in number theory.

Like Noether, Hasse rarely missed a chance to speak at the annual meetings of the German Mathematical Society (DMV), and both were on the program for the 1924 conference held in Innsbruck. Noether presented the axiomatic basis for factorization of ideals into prime ideals [Noether 1924], a prelude to her classic paper [Noether 1927]. This was a generalization of a property Dedekind had proved for the ring of integers in an algebraic number field (today called "Dedekind rings"). Hasse was in attendance at that lecture, which left a deep impression on him. The following year, at the annual DMV meeting in Danzig, Noether spoke about her new approach to representation theory in the context of the theory of abstract algebras. In the same session, Hasse gave a general survey talk about class field theory, in particular the broad new vistas opened by the recent work of Teiji Takagi. This marks the beginning of Hasse's famous "class fields report."

6.5   Takagi and Class Field Theory

Takagi's life story underscores how even a true mathematical genius may have a difficult time finding his or her own way forward.[21] A stimulating intellectual atmosphere will usually be very helpful, but at times isolation can be even more beneficial, particularly if it frees a person from the influential views of leading authorities. Takagi had already studied Hilbert's *Zahlbericht* in Tokyo when he arrived in Göttingen in 1900 [Sasaki 2002, 240], but his timing was unlucky: Hilbert's research interests were no longer focused on number theory. They met occasionally, but Hilbert was more than a little skeptical when Takagi told him he was working on Kronecker's *Jugendtraum*, one of the early cornerstones of class field theory. After three semesters in Göttingen, Takagi returned to Tokyo and gradually came to feel he had been too quick to accept Hilbert's approach to class fields. Heinrich Weber had developed a more general theory, though it was also more complicated than Hilbert's approach [Edwards 1990].

After the outbreak of the Great War, Takagi lost all possibility of staying in touch with European mathematicians, a circumstance that may have emboldened him to drop Hilbert's approach and take up Weber's more general theory. As he later recalled:

> I was freed from that idea and suspected that every abelian extension might be a class field if the latter is not limited to the unramified case.

---

[21]The story is briefly told in [Yandell 2002, 219–230].

I thought at first that this could not be true. Were it false, the idea
should contain an error and I tried my best to find this error. At that
time I almost suffered from a nervous breakdown. I dreamt often that
I had resolved the question. I woke up and tried to remember my
reasoning but in vain. I tried my utmost to find a counterexample to
the conjecture which seemed all too perfect. Finally I made my theory
confirming this conjecture, but I could not rid myself of the doubt that
it might contain an error which would invalidate the whole theory. I
badly lacked colleagues who could check my work. [Takagi 1935]

After a long struggle, Takagi convinced himself that his theory was sound,
but he still faced the difficult task of convincing leading experts in the mathematical world at large. He set out his case in two long papers, [Takagi 1920] and
[Takagi 1922], both written in German. His first opportunity came in September
1920, when he spoke about his results at the International Congress in Strasbourg.
Again, his timing was unlucky, since German mathematicians were not allowed to
attend due to the post-war boycott implemented by the International Research
Council. Takagi first spent a month in Paris before leaving for Strasbourg to take
part in the congress, where he would speak in French. The session in which he
spoke was chaired by the eminent American number-theorist Leonard E. Dickson
from the University of Chicago. Unfortunately, Takagi's presentation left little
impression on him or the few other mathematicians in the audience who might
have been able to follow him under other circumstances.

Before returning to Japan, Takagi visited Hilbert in Göttingen, but apparently missed the chance to meet Carl Ludwig Siegel, who was then studying under
Edmund Landau. Siegel later received a copy of [Takagi 1920], and in 1922 he
loaned this to Emil Artin, who had come to Göttingen that year. Artin had no
difficulty digesting its 133 pages, and came away duly impressed. The following
year, as a private lecturer in Hamburg, Artin had ample opportunities to meet
with Hasse, who held the same position in Kiel. He urged Hasse to read Takagi's
two papers, both of which he studied carefully. As reported in [Honda 1975], after
reading the first paper, "Hasse was deeply fascinated by its generality, its clearness,
its effective methods, and its wonderful results." The second paper, on reciprocity
laws, inspired him even more.

Hasse soon began preparing a lecture course on Takagi's class field theory,
which he taught in Kiel during the summer semester of 1924. Reinhold Baer, who
later became Hasse's assistant in Halle, wrote up a text based on his notes from
this course. This served as the basis for Hasse's lecture the following year at the
DMV meeting in Danzig as well as his "class fields report" [Hasse 1926]. Regarding
the latter, Erich Bessel-Hagen sent a postcard to Hilbert, dated 17 August 1926,
in which he reported:

A few days ago there appeared ... a report by Hasse on class field theory,
which is written with excellent clarity. The design of the whole theory
is wonderfully uncovered by presenting only the main ideas, while the

proofs are reduced to their skeletons. Reading this article is a real pleasure; now all obstacles are eliminated which may have hampered access to the theory. ... [Frei/Roquette 2008, 16]

6.6   Collaboration with Hasse and Brauer

Not long after this, Emmy Noether began to take a serious interest in these developments, which she recognized were closely connected with the arithmetical properties of hypercomplex number systems. She and Hasse (Fig. 6.2) had already been corresponding since 1925, but in 1927 they began an intense and mutually fruitful collaboration. As Lemmermeyer and Roquette described this:

> ... Hasse and Noether had somewhat different motivations and aims in their mathematical work. Whereas Hasse is remembered for his great concrete results in number theory, Emmy Noether's main claim to fame is not so much the theorems she proved but her methods. ... their letters show [they shared] a mutual understanding on the basic intellectual foundations of mathematical work (if not to say of its "philosophy"). Both profited greatly from their contact. [Lemmermeyer/Roquette 2006, 8]

Noether reacted to Hasse's report [Hasse 1926] with enthusiasm in a letter from 3 January 1927. Soon afterward, he began to explore the theory of algebras and their arithmetics, whereas Emmy turned her gaze toward class field theory. Around the time the Noether-Hasse collaboration was warming up, B.L. van der Waerden returned to Göttingen as Courant's assistant. During 1926/27 he was a Rockefeller fellow in Hamburg – a leading center for modern algebra with Emil Artin, Otto Schreier, and Erich Hecke. Artin had agreed to write a book on modern algebra for Courant's yellow series, beginning with a course on the subject that van der Waerden attended. They planned at first to write the book together, starting from the material van der Waerden extrapolated from his course notes. Before the course had ended, however, Artin dropped out of this project, which would occupy the young Dutchman's attention for another three years, during which time he worked closely with Emmy Noether.

In Göttingen, van der Waerden decided to solidify his knowledge of ideal theory by teaching a course on it. He also attended Noether's lecture course on the theory of algebras, the same topic he had heard before, but this time presented in a new form based on representations of groups and algebras. Van der Waerden took careful notes from these lectures and together they turned these into a readable text. These were then elaborated in her groundbreaking article [Noether 1929a], published in *Mathematische Zeitschrift*.[22]

---

[22]The contents of this paper were summarized by Nathan Jacobson, who noted its importance for representation theory in [Noether 1983, 17–18].

Figure 6.2: Emmy Noether, Helmut Hasse, and an unidentified woman, ca. 1930 (Auguste Dick Papers, 13-1, Austrian Academy of Sciences, Vienna)

In the meantime, Noether had struck up a friendly correspondence with the algebraist Richard Brauer, then a private lecturer in Königsberg. By 1927, he and Noether, working independently, had arrived at the concept of splitting fields for simple algebras. This led to their joint paper [Noether/Brauer 1927], submitted to the Berlin Academy by Brauer's former mentor, Isaai Schur. The background behind this publication sheds considerable light on Noether's role in this story as well as subsequent developments that culminated with the Brauer-Hasse-Noether Theorem [Brauer/Hasse/Noether 1932].[23]

Noether and Brauer picked up their discussion again in September at the annual meeting of the German Mathematical Society held in Bad Kissingen. On that occasion, he showed her a counterexample to one of her claims. They also discussed plans for their joint publication. Noether was eager to bring Hasse into the picture, which eventually led to their three-way collaboration. A few weeks after the Bad Kissingen meeting, Noether sent a postcard to Hasse, posing a question she hoped he could answer. Two days later, on 6 October 1927, Hasse

---

[23]This alphabetical order of names, while today quite standard, has not always been adopted. Some authors also add Abraham Adrian Albert to the list (e.g., [Curtis 2007]), whereas the ordering Hasse-Noether-Brauer appears in [Koreuber 2015]. In support of the latter, we cite the final remark of Nathan Jacobson following his discussion of [Brauer/Hasse/Noether 1932]: "One must conclude that the major share of this achievement should be attributed to Hasse" [Noether 1983, 21].

already wrote back with the answer: "Your conjecture is correct, though it's not a *direct* consequence of my earlier existence theorems. But I prove it with entirely similar methods ..." [Lemmermeyer/Roquette 2006, 72]. Noether then wrote to Brauer, sending him Hasse's letter that solved the question under discussion; she now proposed writing a joint note, to be published together with a note of Hasse. This led to a series of letters within the triangle Brauer-Hasse-Noether, as they discussed various details about the planned notes and possible generalizations. As Peter Roquette commented: "in Hasse's letter of October 1927 [we see] the nucleus of what in 1931 would become the Local-Global Principle for algebras in the Brauer- Hasse-Noether paper [Brauer/Hasse/Noether 1932]" [Roquette 2004, 53].

Noether's evident delight shines through in her letter to Hasse:[24]

> Your proof brought me much joy; so the matter lies somewhat deeper! I was thinking of a publication in the Reports of the Berlin Academy, where up to now pretty much all short communications on representation theory can be found. I sent a 5-6 page note to R. Brauer – Königsberg so that he could add his part; it should appear under our names .... Your proof could then follow immediately afterward as a short note; perhaps with the subtitle "from a letter to E. Noether" so that no textual changes would be needed! What should the main title be and do you agree with my ideas at all? Should the notes be bound together or separately? ...

Hasse decided to rewrite his letter as [Hasse 1927], which appeared immediately after [Noether/Brauer 1927]. The contents of these two papers were perhaps less important than the new dynamic that Emmy Noether had created by bringing Brauer and Hasse closer together. In the years that followed a close cooperation developed between these three talented mathematicians, whose special expertise blended together remarkably well.

Noether and Hasse were again both on the program for the DMV conference in 1929, which convened in Prague. Noether's lecture topic was briefly summarized in [Noether 1929c], where she noted that a detailed account would follow in *Mathematische Annalen*. She had prepared such a manuscript well before the Prague conference, but never found the opportunity to revise it for publication. Her former student, Heinrich Grell, obtained this manuscript or a copy thereof, which he submitted to Hasse after the Second World War had ended. Hasse, as editor of *Crelle*, published this sketch in [Noether 1950]. Nathan Jacobson underscored the originality and importance of this posthumously published paper in [Noether 1983, 15–16].

Hasse's remarkable lecture in Prague, mentioned already in the Preface and in Chapter 2, surely warmed Noether's heart. His message was the one she had long believed in and steadfastly promoted, and he duly noted the leading role she

---

[24]Noether to Hasse, 19 October 1927, [Lemmermeyer/Roquette 2006, 76].

played in making applications of ideals a central feature of modern mathematics. "It is thanks to the rich and beautiful successes of E. Noether and to her tireless enthusiasm in writing and teaching that this concept, along with the concept of fields, today ties together the various areas of algebra into a unified whole through its methodological band" [Hasse 1930, 32–33]. He ended with these telling remarks:

> One must not forget that the algebraic method is only a *method*, and thus to apply it one needs *substance*. I mean by that: ideal theory, for example, would have never come about of its own accord through an interest in the definition of an ideal, but rather it required the concrete problems posed by algebraic number theory. And so it is with all of modern algebra. This is why abstract algebra in the long run cannot exist apart from concrete mathematical theories any more than the latter can assert themselves in a lasting way without the systematizing and energizing effects of abstract algebra. [Hasse 1930, 34]

Following Hasse's impassioned performance, Oskar Perron, van der Waerden, and Ernst Zermelo offered comments. Zermelo recommended dropping the adjective "abstract" in favor of "general" field or group theory, in particular because some representatives of the older tradition use the former term pejoratively.[25] Considering that Zermelo's comment was made in the year 1929, it should come as no surprise that "abstract algebra" later became a principal target for proponents of "Deutsche Mathematik" during the Nazi years.[26]

6.7 Noether's "Wish List" for Favorite Foreigners

A year had now passed since Noether's arrival in Moscow, where she and Alexandrov hoped to import her new brand of algebra. She wrote him in August 1929 with various local news, in particular about how Hilbert had engaged Stefan Cohn-Vossen to turn his lecture course on "Anschauliche Geometrie" into a book for Springer's yellow series. She knew how enthusiastic Alexandrov was about this informal approach to topics in geometry, and so did Hilbert, who asked Alexandrov if he would write a short appendix on elementary problems in topology. He tried, but found it difficult to write something so compressed, so instead he wrote a short monograph with a preface by Hilbert, entitled *Einfachste Grundbegriffe der Topologie*, which was later published in [Alexandroff 1932a] alongside the volume [Hilbert/Cohn-Vossen 1932].

During the summer of 1929, Alexandrov had struck up a new friendship with Andrey Kolmogorov, another great mathematician with a deep love for nature.[27] Together they took a long boat trip along the Volga, then journeyed further in

---

[25] Cited in *Jahresbericht der Deutschen Mathematiker-Vereinigung* 39: 17.

[26] On "Deutsche Mathematik," see [Mehrtens 1987] and [Segal 2003, 334–417].

[27] For Kolmogorov's recollections of these times together with Alexandrov, see [Shiryaev et al. 2000, 145–157].

the Caucuses and Crimea as well as the south of France. Alexandrov reported on these adventures in a letter to Emmy Noether that she answered on 13 October 1929:

> Your travel descriptions are fantastic – I have never been so aware of the mixture of East and West in you as in this letter! And basically this summer gave you much more than the usual Göttingen summers! We all expect you next summer as a matter of course, and trust to your skills in overcoming all difficulties.
>
> I'm really looking forward to the "algebraic methods in general topology". The final chapters of the book will probably be in this direction as well? I can easily imagine that the group-theoretic approach ... allows for a much easier transfer to closed sets than using matrix calculations. Is that not so?

After this, she went on to tell Alexandrov something about what had taken place in Prague at the annual conference of the German Mathematical Society, writing about "die Algebra" allegorically (for the missing portion of this letter, see Section 5.2):

> ... Prague showed that there is great interest in topology. ...
>
> Algebra was much more strongly represented; in the evenings she was sitting in thick piles, and there were still some missing, e.g. Krull because of his honeymoon. Artin, on the other hand, appeared with his young wife, a Russian math student whom Artin brought up on ideal theory![28] Talking math with Artin is always a pleasure.
>
> Algebra has altogether been busy getting married this year, starting with Reinhold Baer – the millionaire without millions,[29] as Zermelo has called him .... Van der Waerden was the last; he married in late September and had to start lecturing in early October. In his thoughts, though, he is not yet focused, as he writes.[30] The Galois theory is ready and will be typed next week;[31] so you will get it in time. Tell me how your lectures are going and whether, unlike me, you have already won over the masses in Moscow![32]

Noether also passed on the recent news that Hasse would succeed Hensel in Marburg, so she hoped this would lead to future guest lectures there. Emmy ended her rambling letter with a teasing remark made by Alexandrov's former landlady: "Did I write to you how, at van der Waerden's engagement party, Frau Bruns declared: 'Now Alexandrov cannot be left alone.' So take it to heart!"

---

[28] Artin married Natascha Jasny on August 29, 1929.

[29] Baer's wife, Marianne Kirstein, was the daughter of a (perhaps not so wealthy) bookseller in Leipzig.

[30] In October 1928, van der Waerden was appointed professor of geometry in Groningen; he remained in that position until 1931 when he moved to Leipzig.

[31] Presumably a reference to a text from one of Noether's lectures.

[32] Translated from [Tobies 2003, 104].

Figure 6.3: Emmy Noether lived from 1922 to 1932 in the attic apartment of this house, Friedländerweg 57, Göttingen (Photo form 1966, Auguste Dick Papers, 12-12, Austrian Academy of Sciences, Vienna)

Noether had already written to Hasse one week earlier. She sent him her congratulations, but also let Hasse know that she hoped his appointment would open up new opportunities for two gifted foreigners. One of these was her student Jacob Levitzki, who was born in the Ukraine but grew up in Tel-Aviv after his family left for Palestine in 1913. In 1922, after graduating from the Gymnasium in Tel-Aviv, he took up studies in Göttingen, where he completed his doctorate under Noether and Landau in 1929.[33] In her letter, she expressed hopes that Levitzki might be considered for a position as Hasse's assistant. "He has presently returned home for the first time in seven years; but it is very possible he will come back in the winter, and he will surely come if he has prospects for employment." She then went on to ask:

> The other question is whether it would be possible to put Alexandrov on the list in Halle. I know that Alexandrov very much wishes at some time to come to a German university. And since all his work is published in German journals – or even in America in the German language – perhaps having a foreigner there would not be too difficult, especially since his scientific importance is undisputed. In addition, he is now writing his "Topologie" for the yellow series; and finally, for the two summers when he was a guest lecturer in Göttingen he received a salary from the government, even with tax deductions. Besides his

[33]Noether's report on Levitzki's dissertation is transcribed in [Koreuber 2015, 323–324].

topology course, he is also teaching Galois theory this winter, of course modern; and he always works intensively with his people in seminars. Works by some of his students have already appeared in the *Annalen*, and more will come. You will know that he has complete and perfect command of the German language.

That's the end of my wishes for foreigners![34]

She wrote Hasse again on 13 November apropos Levitzki, but also to ask whether Hasse planned to attend the ceremony for the opening of the new Mathematics Institute in Göttingen, which would take place on December 3. If Levitzki had no chances in Marburg, she planned to buttonhole whomever she might bump into at that gathering. Eventually, Noether wrote to Øystein Ore at Yale University, with whom she co-edited Dedekind's Collected Works [Dedekind 1930–32], and Ore arranged for Levitzki to spend two years at Yale on a Sterling Fellowship. In 1931, he returned to Palestine as professor of mathematics at Hebrew University, where he lay the groundwork for algebraic research in Israel.[35] Hasse's eventual successor in Halle was Heinrich Brandt, who later became a sharp critic of Noether's abstract style.

Alexandrov was probably never seriously considered as a candidate for this position or any other professorship in Germany. He kept coming back to Göttingen, though, joined in May 1930 by Kolmogorov. During that summer semester Pavel Alexandrov offered a lecture course on his new combinatorial theory of dimension, his latest effort in attempting to fill that "deep gorge separating general (set-theoretic) and classical topology" that he described to Hausdorff in his letter from July 4, 1926 (see Section 5.5). Years later, Alexandrov recalled frequent convivial gatherings that summer when he and Kolmogorov socialized with various Göttingen mathematicians. He remembered meeting

> ...most of all with Courant (and I also met Emmy Noether very often), but also with Hilbert and Landau. We were sometimes invited to Landau's house for a (usually very grand) supper, one feature (and the main attraction) of which was an enormous dish of lobster. The guests were asked to demonstrate their skill at eating these arthropods. Kolmogorov was awarded first prize, as he managed his portion of lobster without even once touching it with his hand, and using only knife and fork. [Alexandrov 1979/1980, 327]

---

[34]Noether to Hasse, 7 October 1929, translated from [Lemmermeyer/Roquette 2006, 92].

[35]Edmund Landau gave a speech in Hebrew at the opening ceremony in 1925; it was rumored afterward that he learned his Hebrew from Levitzki; for background see [Corry/Schappacher 2010, 455–456].

6.8 Paul Dubreil and the French Connection

Probably Alexandrov saw Noether less often that summer than he remembered, as she was then teaching in Frankfurt, having switched places with Carl Ludwig Siegel. Her student, Max Deuring, attended Siegel's course on analytic number theory and was assigned the challenging task of writing up a readable manuscript from these lectures using the notes he took. On learning this, Emmy Noether sent him a postcard in which she remarked: "it is very nice that you are preparing the Siegel lectures; then I can read his breakneck proofs in peace in the winter, which I much prefer to hearing them" [Lemmermeyer/Roquette 2006, 92]. Siegel was a leading authority on analytic number theory and, in later years at least, a sharp critic of abstract algebra.

During her summer in Frankfurt, Alexandrov did come over to visit Noether and to speak in the mathematics seminar. One of those who heard him lecture was the young French mathematician Paul Dubreil, who also recalled memorable lectures delivered by two other guests.[36] Dubreil was an important link in the chain connecting Paris and Göttingen, but especially for the "French connection" running from Emmy Noether to the founding figures of the Bourbaki group. His recollections of his years abroad contain numerous interesting anecdotes that nicely capture the atmosphere from that time (see [Dubreil 1982] and [Dubreil 1983]).

In 1929/30 Dubreil was a Rockefeller fellow studying with Artin in Hamburg, one of five locales he would visit over the course of 15 months.[37] During parts of this journey he was accompanied by Marie-Louise Jacotin, a fellow normalien whom he married in June 1930. Before their marriage, she had a scholarship from the École Normale Supérieure that took her to Oslo, where she worked on fluid dynamics under Vilhelm Bjerknes. In Hamburg, Dubreil was swept into Artin's world, which centered around informal group discussions that took place both before and after Artin's lecture course.

In February 1930, Emmy Noether came to Hamburg, having received an invitation from the local Philosophical Society. Dubreil got to meet her over lunch following Artin's lecture, and that evening he went to hear her speak in the university's aula, where she addressed a large audience composed of philosophers and mathematicians. In a letter to Marie-Louise Jacotin, he afterward conveyed his "total disappointment":

> I saw a short, corpulent woman with a ruddy complexion and dressed without any elegance. Obviously very intelligent, she spoke non-stop, very quickly, in a jerky manner. She fell headlong into the trap that this conference was for philosophers and mathematicians: incomprehensible to the first, she was breaking open doors for the others. And all in a big mess: definition of a field by the usual system of axioms, equivalence classes, isomorphisms, theory of ideals with Teilerkettensatz

---

[36]Louis Mordell spoke on number theory and Wolfgang Krull on non-noetherian rings.

[37]The other four were Groningen (van der Waerden), Frankfurt, Rome, and Göttingen.

(axiom ... now called the ascending chain condition), return to fields to say a few words about real fields, no doubt to pay homage to Artin. [Dubreil 1983, 64]

He later came to realize, after he got to know Noether personally, that this talk was meant to illustrate two principles central to her work: the construction of concepts (*Begriffsbildung*) and the formulation of increasingly strong axiom systems within the same theory, thereby yielding more and more precise results.

The day Dubreil left Hamburg, he went to Artin's office to say good-bye and thank him for all his help. This prompted a quite unexpected and memorable exchange. "Oh no," Artin responded, "I have helped you so little in your personal work!" Dubreil: "But, Herr Professor, I had to do my thesis on my own!" Artin then insisted that he tell him about his work dealing with a rigorous general treatment of the "multiplicity of (Max) Noether." It took him almost an hour to explain this, and then to finally say good-bye [Dubreil 1983, 65].

Dubreil then went on to Frankfurt, where Noether was teaching during the summer semester of 1930. After greeting him, he was in for his next surprise. "By the way," she said, "you are speaking in the seminar next week." Dubreil stared at her, taken aback. "Yes, about your results on the multiplicity of Noether" she added. He paused to ask: "And, in German?" "Yes, yes of course!" So Dubreil got straight to work, feeling a little comforted that he had managed to give Artin an improvised explanation of his work before his departure from Hamburg. Moreover, the relaxed atmosphere in Frankfurt appealed to him right from the start. He got to know Wilhelm Magnus and Ruth Moufang and generally felt "supported by a kind of sympathetic prejudice that seemed to float in the air!" After his presentation, Dubreil remembered Emmy Noether's comment: he had treated a problem of her father with the methods of his daughter. The weather that day was splendid, so the group made an *Ausflug* to the Taunus, where Dehn loved to hike. They probably stopped at the Fuchstanz restaurant for cake and coffee; perhaps this was the day Dehn's daughter Maria remembered, when Emmy ordered a second piece of Torte, laughing that it couldn't hurt her figure anyway [Dubreil 1983, 66].

Like nearly everyone else, Paul Dubreil found Noether's lectures difficult to follow; she spoke very fast and her presentations were untidy, so he had his troubles trying to take notes. Still, she frequently returned to important concepts, and so he eventually began to feel at home with her way of thinking. He recalled how one day, though, he was having difficulty with a certain assertion which seemed to him unsubstantiated by her methods, but which he could prove without difficulty using matrices. So after the lecture, he showed her his calculation – knowing full well her strong aversion to matrix methods. After a few moments reflection and with a bit of skillful juggling with her modules, she showed him how the matter was all very clear [Dubreil 1983, 67–68].

Chapter 7

Emmy Noether's Triumphal Years

7.1 The Marburg "Schiefkongress"

When Emmy Noether returned from the September 1929 conference in Prague – where she and Hasse surely spoke about their mutual mathematical interests – she belatedly answered a postcard he had sent here. He was interested to find out what she knew about the relationship between hypercomplex algebra and class field theory. She began by saying, "not much and mainly just formal," and then proceeded to write a long letter describing eight points and then ending abruptly: "Now make what you want of these fantasies" [Lemmermeyer/Roquette 2006, 90]. By this time, Noether's and Hasse's research interests were beginning to converge. After receiving his manuscript for [Hasse 1931], she wrote back 25 June 1930 to express the great pleasure it gave her, but also to point out connections with a number of other recent works. Noether served as an unofficial editor for *Mathematische Annalen*, so her remarks were aimed at improving the text before it appeared in print.

Toward the end of 1930, Hasse formulated some far-reaching conjectures on simple algebras over number fields, including the assertion that they are all cyclic. When he sent these to Noether, her first reaction was disbelief; in fact, when she responded on 19 December, she thought that a recent paper by A.A. Albert's provided a counter-example. "Yes, it is a shame," she wrote, "that all your beautiful conjectures are merely floating in the air and not standing with firm feet on the ground: because some of them – how many I cannot oversee – crash beyond saving due to counterexamples in a completely new American work by Albert [Albert 1930] ..." [Lemmermeyer/Roquette 2006, 99]. Only five days later, however, on December 24, she wrote again with a *pater peccavi* (Father, I have sinned against heaven and before you). "Your castles in the air (*Luftschlösser*) have not fallen at all," she wrote, because what she had originally read into [Albert 1930] was, in fact, pretty much the opposite of what he had proved. Hasse

D. E. Rowe, M. Koreuber, *Proving It Her Way*, https://doi.org/10.1007/978-3-030-62811-6_7

had sent her a counterexample that made this clear to her. She then proceeded
to summarize the results in three of Albert's recent papers, and then concluded
by sending "good wishes for Christmas and New Year, for people and skew fields"
[Lemmermeyer/Roquette 2006, 102].

In January 1931, Heinrich Brandt, Hasse's successor in Halle, invited Noether
to give a talk there. She gladly accepted, proposing to speak about the relations
between hypercomplex systems and commutative algebra. Since she realized that
such a topic could not be easily digested in a single two-hour lecture, she proposed
to hold a follow-up meeting the next day, "a kind of colloquium, where I could give
explanations, or possibly elaborate on the foundations, or similar, more elemental
things" [Jentsch 1986, 6]. It was probably during her visit in Halle that Noether
met the talented young French mathematician Jacques Herbrand for the first time.
He was then on a Rockefeller fellowship in Berlin, and she very likely informed
him in advance about her lecture in Halle.

Hasse sent Noether another postcard with queries in early February 1931,
and again she answered him with a long letter that began in a humorous vein.[1]
She had lots of hypercomplex things to tell him – more perhaps than he really
cared to know. A matter of special interest concerned plans for a forthcoming
meeting in Marburg, the so-called "Schiefkongress" that would take place there
from 26–28 February. She and Hasse had surely discussed this already on January
13, when Hasse came to deliver a lecture before the Göttingen Mathematical
Society. Its title was simply "On Skew Fields (Schiefkörper)," which gave him the
opportunity to present his various conjectures, in particular that central simple
algebras over number fields are cyclic. Emmy had at first been skeptical regarding
these conjectures, but she quickly became convinced that Hasse's hunches were
probably correct; so they both set off in a quest to prove them. This would
culminate near the end of the year when they and Richard Brauer proved the
Brauer-Hasse-Noether theorem [Brauer/Hasse/Noether 1932] (see Section 7.3).

Emmy was not quite a one-woman organizing committee, but she gave Hasse
a clear idea of what she had in mind for the forthcoming "Schiefkongress" in
Marburg, where she would speak about hypercomplex structure theorems with
applications to number theory. She was hoping to have something new to say
by then about a theory of conductors that generalized Artin's version, but if
that was not quite ripe for public display she would talk about the applications
Hasse already knew. She also recommended the following order for the first four
lectures: Brauer, Noether, Deuring, Hasse. Her reasoning for this was that such an
arrangement would enable speakers to take full advantage of the preceding talks,
whereas she imagined the other presentations would be independent contributions.
Noether also asked Hasse to invite Jacques Herbrand, who was then in Berlin
working with John von Neumann. She noted that he was on the first leg of his
Rockefeller fellowship, which would later take him to Artin in Hamburg and finally

---

[1] Noether to Hasse, 8 February 1931, [Lemmermeyer/Roquette 2006, 103–104].

Figure 7.1: Marburg "Schiefkongress," 26–28 February 1931: Back, l. to r.: Ralph Archibald, Max Deuring, Hans Fitting, Gottfried Köthe, Heinrich Brandt; Front, l. to r.: Ernst Zermelo (?), Heinrich Grell, Emmy Noether, Jacques Herbrand, Richard Brauer (Auguste Dick Papers, 12-4, Austrian Academy of Sciences, Vienna)

to Noether in Göttingen. She had already met him in Halle and reported that he had understood her latest ideas better than anyone else there.

Calling this gathering – really only a small workshop – a "congress" was, of course, meant as a joke. For insiders, skewed fields ("Schiefkörper") stood at the center of the "Schiefkongress," for the rest of the German-speaking world, this would have sounded like a big conference that went badly (schiefgehen). Despite its small scale, during this era mathematics meetings were still quite rare, which explains why a short report about this one was published in the journal of the German Mathematical Society. Hasse ultimately arranged the program differently than Noether had in mind. He himself gave the first and only talk on Thursday afternoon, February 26, when his topic was "Dickson skew fields of prime degree." On Friday, Emmy Noether was one of five speakers, and on Saturday there were just two talks.

7.2   Rockefeller and the IEB Program

Jacques Herbrand was invited to attend the Marburg workshop, as can be seen from the group photo (Fig. 7.1). Later that summer he would present two talks in Emmy Noether's seminar, which was attended at this time by three other mathematicians from France. We have already alluded to this "French connection" in Section 6.8, following the vivid recollections of Paul Dubreil [Dubreil 1982, Dubreil 1983]. More familiar still are the ties linking Noether to the Bourbaki group, founded in 1935, the year of her death. As mentioned earlier, two of Bourbaki's founding members, André Weil and Claude Chevalley, had attended her lectures in Göttingen.

Weil was an especially avid traveler in his youth [Weil 1992, 45–60]. He came to Göttingen on a Rockefeller fellowship in 1926/27, during which time he interacted with many people, including Courant, Noether, van der Waerden, and Alexandrov. Before coming to Göttingen, Weil had studied in Rome with Volterra, Severi, and Zariski. His Rockefeller stipend first brought him to Berlin, where he met Heinz Hopf and Erhard Schmidt, but he especially liked to visit Frankfurt, home to Carl Ludwig Siegel, Max Dehn, and Ernst Hellinger. These were well-known, in some cases even famous names, but the mathematical world of the 1920s was still very small. Many of the talented young post-docs on IEB fellowships spent at least part of their time in Göttingen, Rome, Princeton, or Paris.

The main purpose of the IEB program, which was financed by the Rockefeller Foundation, was to stimulate high-level research by promoting international ties and exchanges. Although the outcomes varied a good deal, within the realm of mathematics one can point to several cases – like those of van der Waerden, Alexandrov, and Dirk Struik (see 9.3) – where the IEB helped to launch successful careers. It clearly helped Weil as an itinerant mathematician to connect with a host of interesting personalities, and he snatched up new ideas wherever he went. During his stay in Rome, he heard a lecture by a Rockefeller fellow from the United States, a no longer young woman named Mayme Logsdon. She was an algebraist who had studied under L.E. Dickson, and is remembered today as being the first woman to receive tenure from the mathematics department at the University of Chicago. It was from her that Weil learned of Louis Mordell's article from 1922, which answered Poincaré's query about the rank of the group of rational points on an elliptic curve.[2]

One of Weil's friends from his student days at the École normale in Paris was Paul Dubreil, who was working on his doctorate when Weil returned from

---

[2]Among the 130 mathematicians on Reinhard Siegmund-Schultze's list of recipients of IEB/RF fellowships, Logsdon was one of four women [Siegmund-Schultze 2001, 124–125; 288–301]. Weil later mentioned this incident only in passing; he could no longer remember the name of the American woman who cited Mordell's paper in a paper she distributed [Weil 1992, 49]. Weil would afterward generalize this result by proving what is today called the Mordell-Weil theorem, which opened the way to a new modern discipline: arithmetic algebraic geometry.

Göttingen. Dubreil was then unknowingly following in van der Waerden's tracks, trying to prove a general form of Max Noether's $F = Af + B\phi$ theorem, and he felt seriously stuck. He later recalled what happened some time in 1927:

> These were my tribulations when I had a conversation with my friend André Weil, who had just returned from a visit to Göttingen (...as a Rockefeller Fellow) where he had met Emmy Noether (daughter of Max Noether) and van der Waerden. He pointed out to me that I was immersed, without knowing it, in the theory of ideals of polynomials, and he advised me to read van der Waerden's "Zur Nullstellentheorie der Polynomideale" [van der Waerden 1927] and introduced me to the fundamental work of Emmy Noether and of Wolfgang Krull. Reading these works, which were clear and rich in new ideas, gave me enthusiasm. In July 1928, my thesis was almost finished .... [Dubreil 1982, 79]

For van der Waerden, if he ever heard the story, this would surely have struck him as déjà vu, except that this time it took place in Paris, rather than in Göttingen when he spoke with Noether. Mathematical ideas nearly always spread fastest when traveling by word of mouth.

Having recounted Paul Dubreil's encounters with Emmy Noether in Hamburg and Frankfurt in Section 6.8, we pick up the thread of that story, which eventually led him and Marie-Louise Jacotin-Dubreil to Göttingen. Dubreil defended his dissertation in October 1930 in Paris, after which he and his wife traveled to Rome. He had received an extension of his Rockefeller fellowship to work with Federigo Enriques, while Jacotin-Dubreil continued her work on hydrodynamics with Tullio Levi-Civita. His recollections of these months in Rome bring out very clearly that both Enriques and Severi were keen to exploit the algebraic results of Noether and van der Waerden in the classical setting of complex projective 3-space [Dubreil 1983, 71].

Dubreil's fellowship ended on March 31, but in February Marie-Louise Jacotin-Dubreil received a letter from Bjerknes in Oslo, asking if she would be able to resume work on the French translation of his *Physikalische Hydrodynamik mit Anwendung auf die dynamische Meteorologie*, which was not yet in press (it was published by Springer in 1933). His proposal was particularly attractive because he suggested that they meet in Göttingen, where her husband could resume work under his *Meisterin* Frl. Noether. In fact, once there they worked on this translation together.[3]

By this time, both of them had become seriously interested in abstract algebra, which in her case was quite a leap from hydrodynamics. In Göttingen, though, one could study both at the highest level and so she attended Ludwig Prandtl's course alongside Noether's offerings. Afterward her research remained focused on fluid mechanics, which led to her thesis in 1934 on systems of waves that extended earlier known types. Still, her knowledge of algebra was such that

---

[3]This translation project was never completed, however; [Dubreil 1983, 71–72].

Figure 7.2: A Göttingen *Ausflug* with l. to r.: Hans Heilbronn, Emmy Noether, Marie-Louise Jacotin-Dubreil, Paul Dubreil, and Chiungtze Tsen, Summer Semester 1931 (Auguste Dick Papers, 13-1, Austrian Academy of Sciences, Vienna)

she offered this as a secondary subject for her oral examination. After she and her husband took various academic positions in France, Marie-Louise Jacotin-Dubreil was appointed in 1943 to a full professorship in Poitiers, where she combined her interests in both fields while developing applications of algebra to problems in turbulence and information theory. She and her husband spent many years commuting and arranging their lives until 1955, when both obtained positions in Paris.[4]

That summer in Göttingen the young French couple got to meet an extraordinary array of talented, not to say famous, mathematicians, beginning with Hilbert, who had just retired. His successor, Hermann Weyl, and his wife Helene, were charming hosts, as of course were Richard and Nina Courant. Two of the speakers in the Mathematical Society were Princeton's Solomon Lefschetz and Constantin Carathéodory from Munich, the latter long a fixture in Göttingen,

---

[4]For testimonials of Dubreil-Jacotin's impressive career, see her biography at the MacTutor website.

where he had studied under Hermann Minkowski and then succeeded Felix Klein. After the weekly lectures, the crowd usually gathered for dinner at the Hotel zur Krone, regularly attended by Emmy Noether, the logician Paul Bernays, and Courant's right-hand man, the historian of ancient mathematics Otto Neugebauer. Paul Dubreil remembered meeting Noether's doctoral student, Hans Fitting, Edmund Landau's assistant, Hans Heilbronn (see Fig. 7.2), the number-theorist Kurt Mahler, and the group-theorist Helmut Ulm. He also recalled that Courant's assistant, Hans Lewy, was one of the few who seemed deeply troubled by the recurring displays of Nazis marching in the streets [Dubreil 1983, 73].

The visitors who came to Göttingen that summer were an especially impressive group. One of them, the Finnish analyst Lars Ahlfors, was also on a Rockefeller scholarship. Five years later, at the 1936 International Congress held in Oslo, Ahlfors and Jesse Douglas would become the first recipients of the coveted Fields Medal, occasionally called the "Nobel Prize" for mathematics. Emmy Noether's seminar drew three distinguished young foreigner's, one being Francesco Severi's assistant Beniamino Segre, who had spent the year 1926/27 as an IEB fellow with Élie Cartan in Paris. The other two were former normaliens, both of whom had studied together with Dubreil and Jacotin: André Weil, then returning from India, and Noether's favorite, Jacques Herbrand, whom she had invited to the Marburg workshop that past winter.

Paul Dubreil recalled that Herbrand gave two brilliant presentations. On July 4, 1931, Emmy Noether submitted his latest work on an ideal-theoretic approach to the arithmetic in extensions of a number field for publication in *Mathematischen Annalen* [Herbrand 1932a]. Soon thereafter, he left Göttingen to go mountain climbing in France with two friends. On 29 July 1931 *Le Temps* reported that a man had fallen to his death in the French Alps, and one day later it was confirmed that this was Jacques Herbrand. A clipping from the paper was sent to Dubreil, who immediately informed Emmy Noether. She was deeply disturbed by this news and kept repeating over and again: "such a talent, that's just unthinkable." Emmy Noether then decided to publish excerpts from two letters Herbrand had sent her concerning a problem he formulated and soon thereafter solved; this appeared immediately after his paper in volume 106 [Herbrand 1932b]. Her final paper [Noether 1934] was one of many published in *Actualités scientifiques et industielles* to commemorate his memory.

7.3 Birth of the Brauer-Hasse-Noether Theorem

In the meantime, Noether had been in steady touch with Hasse and Brauer. The Marburg workshop helped to propel Helmut Hasse forward, and already on March 6 he sent out a postcard to the participants informing them that he had proved an extension of the local-global principle to cyclic algebras of any index, not necessarily prime. This was one of the problems he had presented in his talk in Marburg. That eventful summer of 1931 Hasse began to concentrate on his principal conjecture, and wrote to Brauer, Artin, and Noether asking for their thoughts. On July 27 he sent this message to Brauer:

> ...I would like to write to you about the only question which is still open, the question whether all central simple algebras [over number fields] are cyclic. For I believe that this question is now ripe and I would like to present to you the line of attack which I have in mind. [Roquette 2004, 21]

The other three seemed to agree that this question was ready to be attacked, though they offered no very concrete ideas for doing so. In the meantime, Hasse was working on a paper in English that he later submitted for publication in the *Transactions of the American Mathematical Society* [Hasse 1932b]. When he sent Noether some of the results for this paper, she reacted in a postcard from April 12 with a mixture of excitement and advice for a still larger vision:

> I have read your theorems with great enthusiasm, like a thrilling novel; you have really gotten very far! Now ...I wish to have also the reverse: direct hypercomplex foundation of the invariants ...and thus hyper- complex foundation of the reciprocity law! But this may still take some time! [Lemmermeyer/Roquette 2006, 109]

As Roquette pointed out, Hasse became inspired by this vision to reverse his argu- ment, and he would soon succeed in finding what Noether called a "hypercomplex proof of Artin's Reciprocity Law." In this way, "a close connection between the theory of algebras and class field theory became visible" [Roquette 2004, 44].

Hasse had been corresponding with A.A. Albert, then at Columbia Uni- versity, since January 1931. Quite conceivably, Albert suggested to Hasse that he submit a paper to the *Transactions* of the AMS, but even if not, his article [Hasse 1932b] was surely written in the main for Albert's benefit. The middle section of it presents Noether's theory of factor systems for the first time. She gave him permission to make this part of his paper, and in her letter from June 2 she expressed general satisfaction with his handling of her ideas.[5] There she wrote: "I think you have managed to make the thing bite size for the Americans, and also for the Germans, without sacrificing too many concepts." Still, she made a number of critical remarks and probably felt that the overall presentation was not abstract enough to really please her. On the other hand, Hasse was writing "for the Americans," who were less familiar with Noether's abstract ideas. He submitted the paper in late May, and the editors of the *Transactions* then sent it on to Albert for a referee's report [Roquette 2004, 62–67].

Hasse explicitly referred to communication difficulties as one of the motiva- tions for his paper:

---

[5] The term "crossed products" (for Noether's *verschränkte Produkte*) appeared here for the first time, even for her eyes. In a postcard, she asked Hasse whether this was his "English invention" and added: "das Wort ist gut" [Lemmermeyer/Roquette 2006, 109].

The theory of linear algebras has been greatly extended through the work of American mathematicians. Of late, German mathematicians have become active in this theory. In particular, they have succeeded in obtaining some apparently remarkable results by using the theory of algebraic numbers, ideals, and abstract algebra, highly developed in Germany in recent decades. These results do not seem to be as well known in America as they should be on account of their importance. This fact is due, perhaps, to the language difference or to the unavailability of the widely scattered sources. [Hasse 1932b, 171]

The main results and methods in this paper were known to the protagonists long before it appeared in print, but the contact between Albert and Hasse broke off for several months and was only restored when Hasse wrote in October. At that point Hasse had found a proof of his conjecture for the case of an abelian central simple algebra. Albert wrote back on November 6, remarking that he was "very glad to read of such an important result. I consider it as certainly the most important theorem yet obtained for the problem of determining all central division algebras over an algebraic number field" [Roquette 2004, 68]. Hasse, however, only received this important letter about one week after the joint efforts of the three German algebraists cracked the general problem, which occurred before they had time to study what Albert had published in the meantime.

The Brauer-Hasse-Noether theorem, as Hasse's conjecture came to be known, has a precise birth date: it was first proved on November 9, 1931. One day after this, Noether sent a postcard to Hasse: "This is beautiful! And completely unexpected to me, notwithstanding that the last argument, due to Brauer, is quite trivial (Every prime number dividing the index is also a divisor of the exponent)" [Roquette 2004, 8]. Only two days before, Noether had sent Hasse a long letter containing her ideas for improving his paper on the abelian case, which she was able to extend (to the case where the finite group of the required extension field only needed to be solvable). She called her new results "trivializations and generalizations" of those in his manuscript, and then summarized these in the form of a reduction theorem with a simple proof, followed by five easy consequences; she then proposed that he append this to his text [Lemmermeyer/Roquette 2006, 124–126]. When Hasse saw that Noether had now extended his original theorem to the case of solvable groups, he remembered that Brauer had written him earlier that the general case can be reduced to a p-group using Sylow theory, and since these are solvable, Hasse saw how to piece together a proof of the theorem in three steps.

Hasse had been pursuing a different line of argument, which he communicated to Brauer in a 10-page letter. This ends with a description of the roadblock he had hit: "I have to admit that here I am at the end of my skills and I put all my hope in yours" [Roquette 2004, 9]. Only two days later, though, having received Noether's letter, he sent off a postcard to Brauer with the good news: "Just now I received a letter from Emmy which takes care of the whole question ...." Brauer then wrote back to say:

> It is very nice that the problem of cyclicity is now solved! Just today
> I had meant to write you and to inform you in detail about Emmy's
> method; but I have to admit that I feared to make a silly mistake
> because I had the feeling that the thing was too simple. I just wanted
> to ask you about it, but now this is unnecessary. By the way, right from
> the beginning it was clear to me that with your reduction, the essential
> work had been done already. [Roquette 2004, 9].

On November 9, Hasse wrote up a first draft of the proof and sent this to Emmy Noether, who responded the next day via the postcard cited above. In it, she already pointed out how the proof could be simplified further still, a true work in progress!

The final breakthrough that led to the Brauer-Hasse-Noether theorem thus came in early November 1931. This arose, however, from a long and complex intellectual struggle that also involved the birth of local class field theory. This larger context reflected Noether's ongoing interest in non-commutative algebras as tools for both number theory and representation theory. Hasse took charge of writing [Brauer/Hasse/Noether 1932], and he did so essentially by following the historical course of events rather than giving a more conventional and systematic presentation. The proof then involved three reductive steps, the first accomplished by Hasse, then came Brauer's contribution, and finally Noether's.

This paper was also written in great haste, in just two or three days between receipt of a postcard from Emmy Noether, dated November 8, and submission of the manuscript on November 11. This quick rush into print had nothing to do with securing priority, though Hasse knew that Albert was working in the same direction. As co-editor of *Crelle* alongside his former mentor Kurt Hensel, Hasse was preparing a special Festband issue to honor Hensel on his 70th birthday, and this important new theorem made the perfect present for that occasion. Indeed, the first copy was already off the press in time for Hasse to present it to him on December 29, 1931, when Hensel turned 70 [Roquette 2004, 7]. As we shall see in Section 7.5, a similar gift that Courant and Springer had planned to give Hilbert on his 70th birthday failed to be ready on time.

The Brauer-Hasse-Noether theorem was important not only for the theory of algebras; it also pointed the way to a broad new approach to class field theory much as Emmy Noether had long imagined. Her vision for such a theory was also shared by Emil Artin (Fig. 7.3), who sent this reaction to Hasse when he learned of the breakthrough: "... You cannot imagine how ever so pleased I was about the proof, finally successful, for the cyclic systems. This is the greatest advance in number theory of the last years. My heartfelt congratulations for your proof" [Roquette 2004, 6]. Perhaps even more important than the theorem itself, though, were the methods used to prove it, in particular Hasse's local-global principle, to which he indirectly alluded in the introduction:

Figure 7.3: Helmut Hasse and Emil Artin, early 1930s (Courtesy of MFO, Oberwolfach Research Institute for Mathematics)

At last through our joint endeavors we have finally succeeded in proving the following theorem, which is of fundamental importance for the structure theory of algebras over number fields as well as beyond:

Main Theorem. Every central division algebra over a number field is cyclic (or, as is also said, of Dickson type).

It gives us special pleasure to dedicate this result, essentially due to the p-adic method, to the founder of that method, Kurt Hensel, on the occasion of his 70th birthday. [Brauer/Hasse/Noether 1932, 399]

This dedication prompted a brief, but telling remark from Noether, who in a letter to Hasse from November 12 wrote: "Of course I agree with the bow to Hensel. My methods are really methods of working and of thinking; which is why they have crept in everywhere anonymously" [Lemmermeyer/Roquette 2006, 131].

Emmy Noether was quite unhappy with Hasse's style of presentation, but she of course recognized the urgency of the situation. She, too, had written a paper ([Noether 1932a]) for the Hensel Festband, a major undertaking that involved papers from many eminent mathematicians. In a letter from 12 November, she urged Hasse to clarify that he was the actual author:

This is now very nice and extremely convenient for us that we did not have to trouble with the text! But I think you should state in a footnote that you wrote the text – even if one can recognize your style;

this, if only because we get a "bow" in the footnotes and you do not!
...
     Then I would like to have some more precise "historical" information.

What she meant by that related to certain remarks in the text that clouded the picture of their respective contributions. These were easily repaired, of course, but what bothered Emmy most could not be. She complained about the unsystematic form of presentation, which Hasse then defended by pointing to the authority of Otto Toeplitz, who strongly advocated a "genetic" approach as didactically superior. Noether remained unconvinced: "For me personally, the converse [form], which I did not understand, produced the opposite of Toeplitz's joy; I only refrained from a 'motion to systematize' because of the time urgency ...."[6] She insisted, however, that Hasse add a footnote stating that the presentation follows the order of discovery rather than a systematic ordering of the results.

(a) Noether, flanked by Artin and Köthe   (b) In front of the Mathematics Institute

Figure 7.4: Emil Artin with Emmy Noether and Gottfried Köthe in Göttingen, 29 February to 2 March, 1932 (Photos by Natascha Artin, Courtesy of Tom Artin)

Roquette has suggested, however, that in light of this scurried activity, Hasse was eager to make amends:

> Three months later Hasse seized an opportunity to become reconciled with Emmy Noether by dedicating a new paper [Hasse 1933] to her, on the occasion of her 50th birthday on March 23, 1932. There he deals with the same subject but written more systematically. Those three months had seen a rapid development of the subject; in particular Hasse was now able to give a proof of Artin's Reciprocity Law of class field theory which was based almost entirely on the theory of the Brauer group over a number field. Thereby he could fulfill a desideratum of Emmy Noether, who already one year earlier had asked him to

---

[6] Noether to Hasse, 14 November 1931, [Lemmermeyer/Roquette 2006, 133].

give a hypercomplex foundation of the reciprocity law. [Roquette 2004, 11]

A few weeks before this, Hasse visited Göttingen to attend three lectures on class field theory presented by Emil Artin, a special event organized by Emmy Noether (see Fig. 7.4). Others who attended included Edmund Landau, Gustav Herglotz, and Gottfried Köthe, along with several of Noether's students. One of the latter, Ernst Witt, was most impressed and afterward spent his vacations in Hamburg to deepen his knowledge of the subject, which he later transferred to class theory of function fields [Roquette 2002, 45]. Noether also asked Olga Taussky to write up Artin's lectures, with the thought of afterward making mimeographed copies for distribution. By early 1933, when Taussky was still at work on the *Ausarbeitung*, she learned that Hasse's Marburg lectures on class field theory were being prepared for circulation. This made Taussky wonder whether her project to write up Artin's lectures might be pointless, so she wrote to learn Noether's opinion. Emmy agreed that it was *now* probably superfluous to pursue the idea of preparing an elaborated version of the Artin lectures, gently needling Taussky for not having finished this work the previous summer. Noether nevertheless advised her to contact Hasse's assistent, Wolfgang Franz, in order to find out when he expected the Marburg lectures would be ready.[7] A few months later, Noether learned from Hasse that she and others could expect to receive the text from these lectures by May 1. "I didn't mean to criticize you recently," she now informed Taussky, "but rather only wanted to confirm this fact."[8] Noether would later use Franz's elaboration of Hasse's lectures for her teaching in the United States, whereas Taussky's version of Artin's lectures apparently never circulated, although many years later she published an English translation in [Cohn/Taussky 1978].

Olga Taussky remembered that Emmy had told her how she had just turned 50, but that no one in Göttingen had taken notice of this. She then commented wistfully, "I suppose it is a sign that 50 does not mean old" [Taussky 1981, 84]. Three days after her birthday, Noether wrote to thank Hasse in her usual fashion: "I was terribly delighted!" (Ich habe mich schrecklich gefreut!) ..., she wrote, followed by two pages of detailed comments on his paper. Her subsequent letters to him stayed on message; she was as intent as ever on generalizing class field theory by purely algebraic methods. Yet the tone of her letters underwent a marked change that reveal her heartfelt affection for her younger colleague.

7.4 Editing the Works of Dedekind and Hilbert

Among the many mathematicians whose careers were strongly linked with the work of Felix Klein, none were closer to him than Robert Fricke. Perhaps their

---

[7]Noether to Taussky, 4 February 1933, Papers of John Todd and Olga Taussky-Todd, Box 11, Folder 11, Caltech Archives.

[8]Noether to Taussky, 24 April 1933, Papers of John Todd and Olga Taussky-Todd, Box 11, Folder 11, Caltech Archives.

affinity had something to do with the fact that Fricke was a native of Brunswick, the birthplace of Carl Friedrich Gauss. The famous "Prince of Mathematicians" had first attended the city's Collegium Carolinum, forerunner of the present-day Technical University, before taking up studies in Göttingen. The chair Fricke assumed in 1894 had been occupied by Gauss' last student, Richard Dedekind, himself a native of Brunswick. As Dedekind's successor, Fricke had ample motivation to assume responsibility for editing his collected works, though he clearly recognized that most of his predecessor's creative work fell outside the range of his own mathematical expertise.

Fricke may have long harbored a plan to pursue this project, but in any case, he surely realized that Emmy Noether was the ideal candidate to carry it through. When he approached her remains unclear, but circumstances suggest this may have occurred in 1926, one year after Klein's death. At that time, Øystein Ore was working closely with her in Göttingen, and he, too, had strong interests in Dedekind's works. An agreement was thus reached that all three would serve as editors, with Noether and Ore doing the lion's share of the work; this included undertaking a careful study of Dedekind's previously unpublished papers and assorted scientific correspondence.

Around the time that Noether's work on [Dedekind 1930–32] was winding down, Richard Courant was busy orchestrating a similar project aimed at publishing the works of his former mentor, David Hilbert. Courant and the Berlin publisher, Ferdinand Springer, had already hatched a plan for Volume 1, which would contain Hilbert's works on number theory. Springer promised to visit Göttingen on the 23rd of January 1932 to attend the festivities in celebration of Hilbert's 70th birthday, at which time he would present Hilbert with this volume hot off the press. In the meantime, Courant had engaged Wilhelm Magnus from Frankfurt and Helmut Ulm from Bonn to proofread Hilbert's papers and correct any small mistakes. Neither could claim to be an expert in number theory, but both at least had solid backgrounds in algebra.

Such was the situation in Göttingen in September 1931 when Courant set off to attend the annual meeting of the German Mathematical Society, held that year in Bad Elster. There he met with various colleagues, including Hans Hahn from Vienna, with whom he chatted about the status of the Hilbert project. Hahn told him about a young woman named Olga Taussky, who had done her doctoral work on class field theory under Philipp Furtwängler. Taussky also happened to be present in Bad Elster, where she presented a short lecture in hopes of attracting attention and landing a new job. Knowing this, Hahn introduced her to Courant, who was naturally well aware of Furtwängler's stature as a number theorist. He may even have read the latter's proof of the principal ideal theorem in [Furtwängler 1929], a by now famous result that answered the last remaining open conjecture in Hilbert's paper [Hilbert 1902]. In any event, Courant was eager to have Taussky join his editorial team in Göttingen, no doubt hoping that her expertise would help speed up this work.

A few weeks later, she received a letter from him, inviting her to spend the winter semester in Göttingen. Courant therein mentioned a principal duty as well as possible second task: "firstly, to work very intensively in a collaboration to complete the number theory volume of Hilbert's works, and secondly, to possibly also help a little with routines and operations of the Institute." He might have been more forthcoming about what he meant by "routines and operations," but at least he did not conceal that she would be working under time pressure. "The term starts on November 1," he wrote, and "the Hilbert volume must be ready by mid-January."[9] Never one to be intimidated by hard work, Taussky gladly accepted this offer. At the tender age of 25, she would be editing the papers of the greatest mathematician of the past era, a man she would soon meet personally.

Olga Taussky left behind numerous vivid recollections of her difficult early years in Vienna and Göttingen, including various interesting anecdotes about her encounters with Emmy Noether. She first met her one year earlier at the 1930 DMV meeting in Königsberg (Fig. 7.5), where Taussky gave a short talk about her dissertation, in which she sharpened her mentor's proof of the principal ideal theorem. After she had finished speaking,

> ...Emmy jumped up and made a quite lengthy comment which, unfortunately, I was unable to understand because of insufficient training. However, Hasse understood it and replied to it at some length, and there developed between these two mathematicians some sort of duet which they enjoyed thoroughly. Clearly, Emmy was pleased and I even overheard some nice remarks she made about my lecture. She spoke to me frequently later, but not about the subject of my talk. She was very friendly; so was Hasse. All this was very helpful to me, for prior to my lecture I had been justifiably very nervous, for this was a meeting to which very famous mathematicians came, including even Hilbert. ...
>
> When lunch came, I sat down next to Emmy, to her left. ...Emmy was very busy discussing mathematics with the man on her right and several people across the table. She was having a very good time. She ate her lunch, but gesticulated violently when eating. This kept her left hand busy too, for she spilled her food constantly and wiped it off with her dress, completely unperturbed. [Taussky 1981, 79–80]

Little more than two decades earlier, Emmy Noether had undergone a similar initiation rite when she spoke for the first time at the 1909 meeting held in Salzburg. She understood instinctively how important such an experience can be for a young mathematician's future, and Taussky surely left this conference with a new sense of reassurance. She also probably began to get over some of her ill feelings toward Furtwängler. At any rate, she realized now that she had been lucky to have the chance to work on class field theory, a prestigious area of num-

---

[9]Courant to Taussky, 7 October 1931, cited from the translation in [Goodstein 2020, 682].

Figure 7.5: Emmy Noether on a ship crossing the Baltic to attend the DMV conference in Königsberg, September 1930; photo by Helmut Hasse, courtesy of Peter Roquette

ber theory that stood at the very heart of current interest. Furtwängler had many other doctoral students, but nearly all of them worked on other topics.

Olga Taussky went on to a long and illustrious career, recently described by Judith Goodstein in [Goodstein 2020]. She published some 300 research papers in algebraic number theory and matrix representations in algebra and analysis. During the period 1957 to 1977 she taught at the California Institute of Technology. Soon after her retirement, she recounted her life for the Oral History Project of the Caltech Archives, a text later reprinted in [Taussky-Todd 1985]. This provides many details about her early life in Vienna, including the hardships she faced as a student of Philipp Furtwängler. Her teacher came from Elze, an old town on the Leine River roughly 100 kilometers due north of Göttingen going downstream. Furtwängler's grandfather founded an organ construction company there, a business that his father later took over. The family name was thus well-known in musical circles throughout the region, but attained international fame through another branch, to which the conductor Wilhelm Furtwängler belonged.

Philipp Furtwängler was deeply inspired by Hilbert's work on number theory, but as Olga Taussky reported, her mentor never met him personally. This may well have been the case, even though Furtwängler had studied in Göttingen, where he completed his doctorate under Felix Klein in 1896, one year after Hilbert joined the Göttingen faculty. In any event, his introduction to number theory came via Klein, who developed a geometric approach using lattices, not unlike Hermann Minkowski's better-known theory. Beginning in 1902, however, Furtwängler began spinning out a series of articles in which he took up various ideas and conjectures that Hilbert had published in the late 1890s. Furtwängler's first breakthrough came when he successfully answered a prize question that Hilbert posed as a member of the Göttingen Scientific Society. This question was related to the problem of establishing a reciprocity law analogous to quadratic reciprocity – which dates back to Euler and was first proved by Gauss – but for general algebraic number fields. As Helmut Hasse later emphasized in [Hasse 1932a, 530], Hilbert's work launched a reorientation of number-theoretic research, which would henceforth move to ever-higher spheres of abstraction.

Furtwängler, who like many others learned algebraic number theory by reading Hilbert's *Zahlbericht* [Hilbert 1897], remained at a relatively low level in this process. As his student, Olga Taussky took a similar approach to number theory, which for her was ultimately about numbers, not abstract concepts. By the time she began her studies with him, Furtwängler was paralyzed from the neck down, so he had to lecture from a wheelchair, while one of his assistants wrote the equations on the blackboard. Taussky was called upon to do this at times, a task she remembered as very challenging, although her mentor's lectures were models of clarity; Kurt Gödel – whom Taussky met while attending Moritz Schlick's famous philosophical seminar in Vienna – reputedly called Furtwängler's lectures the best he had ever heard. In other respects, though, she remembered him as a poor adviser, whose lack of support only added to her doubts and suffering. His doctoral students – he had some 60 over the course of his career in Vienna – had to wait in a long line outside Furtwängler's office on such occasions when he happened to make himself available. Olga Taussky felt very disoriented, but kept studying the difficult literature on class field theory, hoping he would soon give her a dissertation problem.

She later recalled how, around this time, Emil Artin found a way to reduce Hilbert's conjecture regarding principal ideals to a problem concerning certain finite non-abelian groups.

> Furtwängler did actually tell me a little about this, but without explanations, and made me almost desperate. In the meantime, he proved Artin's group-theoretic statement to be true and hence solved the principal ideal theorem. This was a tremendous achievement, but the world of mathematics was not very grateful and considered his proof as ugly.

In fact, they had little appreciation for his earlier pioneering work either. In spite of my grievances against him as a teacher, I feel his work deserved better credit. [Taussky-Todd 1985, 316–317]

Taussky found the atmosphere in the middle-sized town of Göttingen vastly different than in Vienna, but unfortunately she had barely any time available to immerse herself in its mathematical life. Not only did she, Magnus, and Ulm have to spend long hours each day working on the number theory volume, she was also burdened with the unpleasant task of correcting papers for the assignments Courant gave to students in his course on differential equations. Nevertheless, she still found time to attend Noether's seminar, to which she had been specially invited. In fact, when she arrived from Vienna, "[Emmy] immediately announced to me proudly that she and [Max] Deuring, her favorite student, had studied class field theory . . . and that she was going to run a seminar on this subject because of my visit" [Taussky 1981, 80–81]. On one occasion in that seminar, Furtwängler's proof of the principal ideal theorem came up. When Noether echoed the opinion Olga had heard all too often – it's so "unattractive" – Taussky blew up at her, calling this an unfair criticism: it had taken decades to resolve Hilbert's conjecture – and, after all, it could have been wrong! "I was amazed by my daring," she later recalled, "and so were the others. However, Emmy was completely calm, and I felt certain, was not in the least angry with me. I recognized for the first time that she was a person who did not mind criticism" [Taussky 1981, 81].

7.5   Olga Taussky on Hilbert's 70th Birthday

Noether was naturally interested to know about the state of progress with the Hilbert edition, while she and Ore continued their work on volume 3 of the Dedekind *Werke*. According to Taussky, Emmy was the only one in Göttingen whom she and her co-workers could consult on technical matters, and they probably did not turn to her often, given that she was far more an algebraist than a number theorist. In any event, if Noether did provide help, it was treated informally, since Courant did not wish for her to be directly involved in the Hilbert editorial project [Rowe 2020a, 5]. Nor did he want his staff of proofreaders to append any editorial remarks to the texts; if alterations were required, these were to be entered without any indication that such a change had been made to the original text. This situation stands in sharp contrast with the Klein edition [Klein 1921–23], even though that project also began just as Felix Klein was approaching age 70. In that case, however, Emmy Noether had been one of several younger mathematicians who worked closely with him to prepare those three volumes. Unlike Klein, Hilbert never enjoyed looking backward. He had no patience for "scholarly studies," even if this only involved supervising such work. So he put Courant in charge of this editorial project, just one of the many tasks the latter managed as Director of the Mathematics Institute.

Part of Courant's success stemmed from his ability to uphold and to capitalize on the traditional prestige of the Göttingen mathematical tradition, which Klein and Hilbert had so long embodied. Emmy Noether proved to be a tremendous asset for Courant's Göttingen, but Olga Taussky realized that many there showed little respect for her achievements, despite the recognition she enjoyed elsewhere. Taussky once heard the distinguished Erlangen algebraist, Wolfgang Krull, emphatically say: "Miss Noether is not only a great mathematician, she is a great German woman!" [Taussky 1981, 84]. Richard Courant was a skillful, if unorthodox manager, but he was far too busy to cultivate close personal relations with Emmy Noether. He was also exceedingly protective of Hilbert's image and knew full well that his former mentor, especially in his old age, demanded unconditional loyalty.

As guardian of the Hilbert edition, Courant aimed to mobilize the support of several leading younger mathematicians, whom he enlisted to write commentaries on Hilbert's contributions to various disciplines. In the end, five essays of varying length were published along with Otto Blumenthal's biographical essay [Blumenthal 1935]: Helmut Hasse on number theory (Band 1); B.L. van der Waerden on algebra and Arnold Schmidt on foundations of geometry (Band 2); Ernst Hellinger on integral equations and Paul Bernays on foundations of arithmetic (Band 3). One might well wonder why Courant did not ask Emmy Noether to write an essay on Hilbert's contributions to invariant theory. Perhaps he did and she deferred to van der Waerden, who merely summarized the significance of Hilbert's work for subsequent developments, in particular Noether's important studies, in two pages.

Yet Taussky-Todd's passing comment that Courant did not want Noether involved with the Hilbert edition was surely correct, and although there may have been a number of reasons for this, the most obvious problem stemmed from an evident conflict of interest. Emmy Noether's outspoken enthusiasm for Dedekind's work, which only grew with each volume she edited, was by this time known throughout the Göttingen institute. As Taussky recalled, "[s]he came to appreciate Dedekind's work to the utmost, and found many sources of later achievements already in Dedekind. Occasionally she annoyed even her friends by this attitude. She managed to rename the Hilbert subgroups the Hilbert-Dedekind subgroups" [Taussky 1981, 81]. Courant very likely heard about, or may have even read, Noether's commentary on [Dedekind 1894b] and [Dedekind 1895], where she sided entirely with Dedekind's position in his rebuttal of Hurwitz's approach to ideal theory.[10] This alone would surely have disqualified her in Courant's eyes, since he knew that Hilbert had been involved in this controversy and stood firmly on the side of Hurwitz. In any case, Courant had but one thought in mind: Taussky and co. needed to finish proofreading the papers for volume 1, which included Hilbert's *Zahlbericht*, by early January; that gave them less than three months.

---

[10]Her commentary began: "The latest developments have fully and completely affirmed the correctness of Dedekind's views both with respect to the definition of ideal and divisibility as well as the foundation of the decomposition theorem" [Dedekind 1930–32, 2: 58].

The celebration of Hilbert's 70th birthday would be a major event in Göttingen, but as that time neared it became increasingly clear that Courant's deadline would be very difficult to meet.

In later years, Olga Taussky told various versions of this story, one of which Constance Reid retold as follows:

> In the course of her work on Hilbert's papers, Fräulein Taussky was astonished to discover many errors. These were not typographical errors. Perhaps the bound of a function would be wrongly computed, a theorem incorrectly stated, a step omitted in a proof, or an entire proof necessary to the argument dismissed as "easily seen" when it was not. Although she recognized that, because of Hilbert's powerful mathematical intuition, the errors had not affected the ultimate results, she felt that they should be corrected in his collected works. She was encouraged in this by Emmy Noether, who was editing the work of Dedekind and frequently announced, loudly, that no one would be able to find a single error – "even with a magnifying glass!" [Reid 1970, 200].

Although she was the source for this story, Olga Taussky-Todd was not at all pleased by Reid's account of it, as the latter learned some years later.[11] In view of these misgivings, Reid revised the above passage in order to satisfy her concerns. Whether by design or mere accident (due to lack of space on the page), Emmy Noether was no longer mentioned at all in [Reid 1996, 200]. Yet she clearly was interested in the Hilbert project, at least from afar. Taussky-Todd, on the other hand, was eager to set the record straight about certain "false rumors" that spread around Göttingen at that time. People were saying that "our work concerned small defects only and delayed the publication unnecessarily, and this was even mentioned in C. Reid's volume on Hilbert" [Taussky 1981, 82] (thanks to Taussky-Todd's input).

To refute these grumblings, which she clearly attributed to Courant's influence, Taussky-Todd tried to emphasize that the delay had been due to serious mistakes, not just a few typos. Although she could no longer remember precisely what these were, she did recall speaking to Emmy Noether about the difficulties she had encountered: "Emmy was truly amazed when I told her that Hilbert's work contained many errors. She said that Dedekind never made any errors. Hilbert's errors were on all levels" [Taussky 1981, 81]. This invidious comparison with Dedekind, who was without question an exceedingly exacting thinker and precise writer, rings rather hollow without any evidence to support the claim that Hilbert's papers were full of *substantive* errors. If anyone really cared to find out, they could run a comparative analysis of the original papers with the printed edition. Short of that, one can easily imagine that back in 1931 Olga Taussky experienced a serious conflict of interest when she read and then edited

---

[11] For details about their ensuing correspondence, see [Rowe 2020a].

[Hilbert 1902], with its various claims that her former teacher had struggled so mightily to prove.

Although Furtwängler's papers were lauded by cognoscenti – Edmund Landau and Hilbert's student Rudolf Fueter reviewed them for the *Jahrbuch über die Fortschritte der Mathematik* – Olga clearly felt they were little appreciated by others. Moreover, since all his works postdated Hilbert's publications on number theory, one finds no mention of them in [Hilbert 1932], except for Hasse's closing essay [Hasse 1932a]. Taussky did, however, take the opportunity to add an editorial footnote on p. 506, in which she wrote somewhat vaguely that one of Hilbert's conjectures in [Hilbert 1902] was incorrect as stated. She knew this, of course, because Furtwängler had corrected that mistake.

Taussky-Todd also provided details about the celebration in the Hilberts' home, which Reid described as follows:

> On the day of the birthday itself, Ferdinand Springer, who was the publisher of the collected works, came to Göttingen to present personally to Hilbert the special white and gold leather-bound copy of the first volume. The beautiful cover contained, however, not the printed pages, but only the proofs of the pages; for Fräulein Taussky was still not satisfied. Hilbert made no comment on the unfinished nature of the volume. But later, in his presence, Fräulein Taussky declined a certain brand of cigarette as being too strong for her. Somebody said that one really couldn't tell one brand from another. "Aber nein!" Hilbert said. "Fräulein Taussky can tell the difference. She is capable of making the finest, the very finest distinctions." She was not sure, but she thought he was making fun of her for taking so seriously errors which he himself considered unimportant. [Reid 1970, 201].

Olga Taussky-Todd was particularly displeased about the final sentence, which put the whole episode in an ironic light. She later wrote Reid,

> ... there is a real misunderstanding there. Hilbert made his remark by no means in a joking tone, it was quite an unfriendly one and he needled me quite a bit later at that party, too. I was afraid that he had seen that footnote!
>
> You see, I never discussed anything with him, nor was I expected to. He had moved for years through different streams of subjects, finally to logic and had no knowledge, nor even interest in that type of work any longer.[12]

Olga Taussky-Todd's memories of her stay in Göttingen all point in the same direction: this was a most unpleasant time for her. One of the few positive aspects, though, came from the friendship she managed to strike up with Emmy Noether, despite their differences in temperament and mathematical orientation.

---

[12]Taussky-Todd to Reid, 5 December 1977, [Rowe 2020a].

As for her recollections of this early work on the Hilbert edition, she emphasized that a good deal of controversy concerned Hilbert's *Zahlbericht*.[13] Emmy Noether seems to have been one of its critics, if we can judge by a brief comment she made by way of praising Hasse's first report on class field theory: "I'm happy to see that you're bringing Takagi in order. I notice more and more how much your report helps make it easier to penetrate [the theory]; one only has to compare it with Hilbert's; how much unnecessary effort that requires today!"[14] Taussky-Todd wrote that Noether was *not* among those who criticized the *Zahlbericht* when she was proofreading it, but then added that in Bryn Mawr Emmy once burst out against it, claiming that Artin said "it delayed the development of algebraic number theory by decades" [Taussky 1981, 82]. Presumably what Artin meant was Hilbert's class field theory, not the *Zahlbericht*, since Artin's work on general reciprocity laws – his solution of Hilbert's ninth Paris problem – followed in the wake of Takagi's general theory of class fields.

What seems particularly baffling is that Olga Taussky apparently never mentioned [Hasse 1932a], the short essay at the end of the volume. This was the one place in it where the reader was offered a larger overview of Hilbert's work in the light of subsequent developments. Moreover, this brief survey touches on studies by Furtwängler, Takagi, Artin, and by Hasse himself. One year earlier, he used the local-global principle to prove his Norm Theorem for cyclic Galois extensions of an algebraic number field [Hasse 1931]. This theorem not only generalized earlier results of Hilbert and Furtwängler, Hasse also showed how it could be applied to deduce new results on the structure theory of algebras over number fields. In this connection, he did not neglect to mention Emmy Noether's vision for a general class field theory, about which he commented:

> As a result of these applications and by means of the general conceptual structures of E. Noether, this theory appears to open the way conversely to still unexplored important problems in algebraic number theory, namely to generalize class field theory to general relative-Galois number fields. However, this development is still too young to be a subject that could already be reported on here. [Hasse 1932a, 535]

Hasse's survey was written with a good deal of verve and subjective opinion, and Noether clearly read it with real interest when it was still in page proofs. Her reaction, transmitted by Taussky in a letter to Hasse, reflects not only her desire to set the record straight but also her sense of self-irony. This letter was written on 15 March 1932, thus roughly two months after Hilbert's birthday party. Noether had apparently gone to speak with Taussky, Ulm, and Magnus after she noticed a glaring inconsistency in the text. As Olga Taussky put this:

---

[13]Taussky's opinions about this led to a number of bizarre claims published in [Rota 2008]; these were exposed as "fake news" in [Lemmermeyer 2018], though on last viewing of the biographical article for Taussky-Todd in Wikipedia, Rota's book is still cited there.

[14]Noether to Hasse, 17 November 1926, [Lemmermeyer/Roquette 2006, 57].

Professor Noether pointed out to us just now that there seems to be a contradiction in your afterword to Hilbert's works on algebraic number theory in regard to the historical information about the Hilbert-Dedekind theory of Galois number fields, since this speaks of the completely new discoveries of Hilbert, but then also about the earlier studies by Frobenius. ...

Since we will receive another revision, it might still be possible to make some minor changes without causing difficulties with the printing, such as the following ...

Taussky then suggested two minor emendations, which Hasse adopted for the published text. But Hasse also had the pleasure of reading Emmy Noether's handwritten comments, which also appeared in this letter, quickly softening its tone:

The Dedekind edition is to blame for my appearance as a historical complainer! But the letter is not intended to be as official – or as demanding – as it looks! What bothered me, though, was that you present the *connection between Galois theory and ideal theory* as the *fundamentally new* thing with Hilbert, whereas Dedekind had published at least as much about it ... And the density theorem from 1880 (although first published in 1896) is precisely this connection!

Formally, this inaccuracy comes to haunt you in the contradiction pointed out by Miss Taussky!

But by no means do I want to deny that it was only through Hilbert that these things came to life for number theorists – and that in all likelihood Hilbert found everything completely independently.

Apart from that, the others are probably not such hair splitters as I am, in case you want to leave everything as it is!

Kind regards,
Your Emmy Noether.

7.6 Zurich ICM in 1932

In September 1932, Olga Taussky attended the Zurich Congress (Fig. 7.6) with its several highlights, Emmy Noether's plenary lecture [Noether 1932b] being one of them. A few months earlier, on June 3, Noether wrote Hasse about her preparations for this major event:

... while working on my Zurich lecture, I for once read Gauss. It has been claimed that a half-way educated mathematician knows the Gauss principal genus theorem, whereas only exceptional people know class field theory. Whether that's true, I don't know – in my case that knowledge went in the reverse order – but at least I learned a lot from Gauss

in terms of comprehension .... [T]he transition from my version to the
Gaussian is direct ....What I'm doing is generalizing the definition
of genera by means of characters. [Lemmermeyer/Roquette 2006, 165].

Figure 7.6: Group Picture of Participants at the 1932 ICM in Zurich (Auguste
Dick Papers, 13-1, Austrian Academy of Sciences, Vienna)

It was in Zurich that Noether first presented her version of the principal genus
theorem for number fields, an idea she first developed in early 1932; she published
a complete proof in [Noether 1933b]. Her letter to Hasse is particularly interesting,
not to say surprising, in view of Dedekind's great appreciation for Gauss's theory,
as Noether surely knew from reading his Festschrift contribution [Dedekind 1877].
She, in fact, wrote the notes that appear in [Dedekind 1930–32, 1: 158], but with
no mention of Gauss whatsoever. Her comments begin with reference to a letter
Dedekind wrote to Frobenius in 1883, which reveals that [Dedekind 1877] was
conceived in the context of a general theory of reciprocity laws. After noting that
Dedekind's theory resembles modern class field theory, as developed by Heinrich
Weber, she pointed to Takagi's work as the next stage in these developments. Her
concluding remarks then read: "For ideal theory the significance of this study is
that it treats for the first time relationships between ideals in different rings by
means of mappings of intersection- and extension ideals. The concepts developed
here can be extended to general rings ...." Noether then referred the reader to

the dissertation written by her student Heinrich Grell.[15] In summary, it would seem very clear that Emmy Noether read Dedekind à la Bourbaki, i.e. looking forward and not backward.

The long, convoluted story of the principal genus theorem from Gauss to Noether can be read in [Lemmermeyer 2007]; apparently few picked up later where she had left off. Noether's version was based on an important innovation, however, her theory of crossed products and factor systems, which Hasse had introduced to American mathematicians in his paper [Hasse 1932b]. Noether had developed these ideas as part of her lecture course from the summer semester of 1929. Max Deuring wrote this up, and copies of his *Ausarbeitung* circulated within Noether's network, which is how Hasse came to learn about it. In [Noether 1933a, 644], she noted that readers could refer to Hasse's paper for the theory or another version closer to her lectures in Deuring's forthcoming report, which appeared two years later in [Deuring 1935]. Nathan Jacobson, as editor of Noether's collected papers, decided to include Noether's original lectures, as prepared by Deuring, which were published in [Noether 1983, 711–763].

Another highlight of the ICM was a private party thrown by Teiji Takagi at the Hotel Eden on the Zurichsee, one of those rare occasions when a whole group of leading experts on class field theory could meet face-to-face. A number of younger Japanese mathematicians attended, including Takagi's star student, Shokichi Iyanaga. Among the Europeans were Emmy Noether and several close associates of her school: Taussky, Hasse, and van der Waerden. Others included the Ukrainian Nikolai Chebotaryev, whose density theorem gave Emil Artin the tool he needed to prove his reciprocity law, and Claude Chevalley, a friend of Iyanaga with whom he studied class field theory under Artin in Hamburg. Chebotaryev had already met Noether and Hasse at the 1925 DMV conference in Danzig, and they soon got into a spirited discussion. The young Japanese watched this display of exuberance by these raucous Europeans with a mixture of amusement and disbelief. Most of all they were struck by the loudest of them all: Emmy Noether. As they were getting ready to depart, she asked the Japanese to show her how to bow properly, a scene that left a lasting memory with some of them [Yandell 2002, 229].

Taussky greatly revered Takagi and even tried to pick up some Japanese so she could speak with him in his native language. He spent five months in Europe, much of it traveling with Iyanaga. During his stay in Vienna, Taussky introduced him to her teacher, Philipp Furtwängler, whose work represented the culmination of Hilbert's vision for class field theory. Takagi's publications, on the other hand, marked the beginning of a new vision that inspired the work of Hasse and Artin. In Hamburg, he had the opportunity to meet with Artin, an encounter very unlike his reunion with Hilbert in Göttingen. "Observing my old master grumbling as if speaking to himself," he later wrote, "I wept in my heart" [Honda 1975, 165].

---

[15][Grell 1927]; Noether's report on Grell's dissertation appears in [Koreuber 2015, 318].

For Takagi, the Zurich Congress marked a highpoint in his life, much as it did for Emmy Noether.

After Olga Taussky returned to Vienna, she remained in contact with Noether, who wrote her a friendly letter on 12 November 1932.[16] Emmy began by informing her that she had just sent off a letter of recommendation in support of Taussky's application for an undisclosed academic position. At this time, Taussky had no income aside from the money she made by tutoring students. As usual, Emmy Noether's letter brought various bits of news, such as the recent arrival of Wolfgang Gröbner, who had just taken his doctorate under Furtwängler in Vienna. He came to spend the year working with Noether and regretted that Taussky was no longer in Göttingen.

Emmy included a few comments about her lectures, which her student Wolfgang Wichmann was writing up for her. Apparently Taussky had plans to visit Göttingen again, as Noether commented that she hoped Wichmann's version would enable her to work through the material. In the meantime, Noether had been forced out of her small apartment in the house at Friedländerweg 57, where she had often entertained in the past. Pavel Alexandrov was familiar with the circumstances behind her expulsion, and he recalled these in his memorial address from 1935:

> She did not hide her sympathy toward our country and its social and governmental structure, despite the fact that such expressions of sympathy were considered shocking and improper by most representatives of Western European academic circles. It went so far that Emmy Noether was literally expelled [from her boarding house] at the insistence of the student boarders, who did not want to live under the same roof as a "pro-Marxist Jewess" – an excellent prologue to the drama that came at the end of her life in Germany." [Alexandroff 1935, 8].

The building in which she formerly lived belonged to a fraternity, the Burschenschaft Thuringia, one of several fraternal organizations in Göttingen, many with a long tradition. The Corps Hannovera Göttingen, a fraternity with compulsory academic fencing (mensur) was particularly proud to count Otto von Bismarck as a former member. Given the ultra-nationalist political orientation of these groups, one might imagine Noether was glad to live somewhere else. She found a new apartment in the large building located at Stegemühlenweg 51, not far from her former dwelling. In her letter to Taussky, she remarked,

> My apartment is also very nice and pleasant with the central heating. Alexandroff has been living with me for 14 days, but he has to return to Russia at the end of November. He would very much like to have the little picture of Veblen, where you cut me and Frau Veblen away; he thinks it's very good, even said it should be enlarged.

---

[16]Papers of John Todd and Olga Taussky-Todd, Box 11, Folder 11, Caltech Archives.

Olga Taussky had occupied an office next to Oswald Veblen's at the Mathematical Institute; she could sometimes hear him practicing his lectures on relativity theory through the wall that separated them [Taussky-Todd 1985, 323]. Two years later, he would help arrange a fellowship for her to spend one year at Bryn Mawr College, where she would rejoin Emmy Noether.

The fact that Pavel Alexandrov was her house guest at the new apartment on Stegemühlenweg explains why he would have known about the circumstances that led her to move out of her old quarters. Little did either of them know that this would be their last time together. In his autobiography, Alexandrov remembered this last stay with Emmy Noether nostalgically:

> I spent my last day in Göttingen making farewell visits. Emmy Noether accompanied me. We went first to the Landau's; then to Hermann Weyl's; then we were invited to the Hilbert's for coffee at 5 o'clock.
>
> I never saw Hilbert or Weyl or Landau again. I was invited to a farewell supper at the Courant's at 8 o'clock. Courant's closest mathematical friends were there – Neugebauer, Friedrichs, Lewy, and, of course, Emmy Noether. My train was to leave at 5 a.m. and it was decided that all the people assembled at Courant's house should spend the whole night with him and that we should then all set out for the station together. After a very long drawn-out supper we had a musical evening, which chiefly consisted of the Schubert trio in E-flat Major. It was played by Stefan Cohn-Vossen (piano), Hans Lewy (violin) and Frau Courant (cello). All three played superbly, with a great uplift. I had always loved this Schubert trio, but after this performance on my farewell night in Göttingen it came to occupy a special place in my appreciation of music and altogether in my consciousness and my life. Finally we went through the dark avenues of Göttingen by night to the station, and I left. I never saw Emmy Noether again, so that this parting was also forever. [Alexandrov 1979/1980, 328–329]

Olga Taussky would, however, get to see Noether again, but not in Göttingen; after a two-year pause their world lines eventually converged during the fall of 1934 in the town of Bryn Mawr, Pennsylvania, a place neither had probably ever heard of before they took flight from Europe.

Chapter 8

Cast out of her Country

8.1 Dark Clouds over Göttingen

When Pavel Alexandrov wrote about his last visit to Göttingen some four decades later, he could hardly look past the traumatic events that were to follow in the wake of his departure. His description of the atmosphere in the town little more than two months before Hitler would come to power set the stage:

> But in November 1932 clouds were already thickening over Germany. Often I was woken in the morning by the sounds of "Deutschland, erwache." This was sung by the young people of the "Hitler-Jugend", as they marched up and down the streets. It was clear that things were about to happen and that it was time for me to go home. The day of my departure finally arrived. As I have said, it was at the very end of November.

Göttingen had long been a stronghold of the Nazi Party in northern Germany, which adds plausibility to these personal recollections. Yet like most such retrospective accounts, this one reflects a highly filtered view of the past, which is nothing unusual, of course, since this is how human memory typically operates. Alexandrov surely did hear the Schubert trio in E-flat Major performed in the Courants' home some hours before his train departed that night. What he just as surely did not experience was a deep foreboding that all this was about to end. For, in fact, he was planning to come back to Göttingen in 1933 along with his friend Kolmogorov, and Emmy Noether was very much looking forward to their future visit. This is what she wrote to Alexandrov on 5 March 1933:

> My dear Alexandrov!
>
> The beginning of the holiday – we're lazing around in March and April – gives me the opportunity to finally answer your lovely Christmas and New Year's letter ....

D. E. Rowe, M. Koreuber, *Proving It Her Way*, https://doi.org/10.1007/978-3-030-62811-6_8

I've lived pretty much from hand to mouth since then, partly for
my lecture course, but even more for a lecture in Marburg, where I
had promised Hasse the newest findings – result, everything else was
left behind. It's again hypercomplex number theory, in the sense of my
Zurich lecture .... We're, of course, still far from a general theory of
non-Abelian fields, but the hypercomplex offers the possibility at least
to formulate questions and conjectures and to create work for years,
even for a whole lot of people!

I'm very happy that everyone in your institute is working so well;
you'll soon be raising a younger topological generation, and since Pontryagin and my other acquaintances already have positions as parents,
you will be able to enjoy your grandchildren while you are young. ("I
hear you feel like a grandmother," Fischer once said to me in regard to
F.K. Schmidt-Erlangen).[1]

It is nice that Kolmogorov will come in the summer and that you
will keep him company here in winter at least. Frau Bruns will be interested to know whether Kolmogorov will stay with her in the summer.
Lately I've had lots of visitors under my Noether roof, but only for short
stays. You can follow that next year by reading the guest book, where
all sorts of mathematical poets have emerged, above all Neugebauer
with his dedication.

These last days van der Waerden was here for the habilitation of
his brother-in-law [Franz] Rellich.[2] V. d. Waerden is now much fresher
than he was not long ago. ... [Tobies 2003, 104–105]

Van der Waerden had left Groningen in 1931 to become professor of geometry
in Leipzig; this was the prestigious chair previously occupied by Felix Klein, Sophus
Lie, and Otto Hölder. Noether probably knew that van der Waerden had applied
for funding from the Rockefeller Foundation in order to spend a semester in Rome
working with Francesco Severi, whom he had met six months earlier at the ICM
in Zurich.[3] She alluded to this when she wrote:

Courant thinks that with time the planned stay in Rome can take
place, just not so quickly. Severi continues meanwhile to make propaganda for Italian algebraic geometry, now even in notes for the *Annalen*![4]

---

[1] Friedrich Karl Schmidt took his doctorate in Freiburg in 1922, formally under Alfred Loewy,
though the topic for his doctoral thesis was suggested by Wolfgang Krull, who also took his degree
under Loewy. Schmidt was also strongly influenced by Helmut Hasse before he habilitated in
Erlangen in 1927. Ernst Fischer was presumably alluding to Noether's connection with Krull,
implying that she could think of herself as Krull's academic mother.

[2] Franz Rellich was a close protégé of Richard Courant; after the war ended, he was appointed
director of the Göttingen Mathematics Institute.

[3] Three days earlier, Courant wrote a letter of support for van der Waerden's application,
transcribed and translated in [Schappacher 2007, 265]. This plan was never realized.

[4] Here Noether seems to have been referring to [Severi 1933], a paper reviewed by Hellmuth
Kneser. The latter commented about it, "general and personal remarks scattered throughout

Courants will drive to Arosa tomorrow with the whole family, including Miss Maier and Rellich; at the end of the month I'll again go to the North Sea for 2-3 weeks. Now that Weyl has finally rejected the offer from America,[5] Courant will be protected for a number of years from the burning question of having to bring a successor to Göttingen. You said it right: there was a mood of disaster over the possibility of needing to find a successor. By the way, this matter caused Weyl to become ill; he has been on leave since Christmas to cure his nerves, even gave up the trip to America he had planned for February. But he plans to be here again in May, which will probably be pleasant for Kolmogorov. . . . [Tobies 2003, 105]

Emmy Noether's letter went on with all kinds of other mathematical news, plus wishes for pleasant skiing weather. But there was not a word about politics or any sense that she and her colleagues might be affected by recent and certainly imminent Nazi policies and threats. Perhaps that was only prudent, but the tone in all her correspondence from this time onward was consistently apolitical. On February 4, she sent a short note to Olga Taussky in which she mentioned how Hasse and Brauer came to speak in her seminar on the same day. This was shortly after Christmas in the dead of winter, but with the obligatory walk to Kerstlingeröder Feld, where Brauer held his lecture under petroleum lights. Or a few months later, after she learned that the government had suspended her from teaching: "I just spent three weeks on the North Sea with wonderful weather; one can then follow contemporary events as an observer."[6]

Hermann Weyl would not have been surprised, since that was how he knew her: "There was nothing rebellious in her nature; she was willing to accept conditions as they were" [Weyl 1935, 435]. In her world, as reflected in these letters, the general mood during the past few months had only been darkened by a seemingly trivial matter, namely, the uncertainty over who might take Weyl's place were he to decide to leave Göttingen. But now that this dark cloud had passed, life could go on as usual. Clearly, Weyl had little more inkling than Noether about what the future held in store; otherwise, he could have saved his nerves, accepted Flexner's offer, and boarded a ship bound for New York in February. If there was anyone who might have sensed trouble, it should have been Courant, who had many real enemies both inside the university and beyond it. He knew this well enough, but he, too, underestimated the danger. So the moment the winter semester ended, he drove off with his family to go skiing in Switzerland.

---

the article impart even to the non-initiated reader a lively impression of the peculiarity and achievements of the author and the Italian school" [Schappacher 2007, 266].

[5]Abraham Flexner, director of the newly founded Institute for Advanced Study in Princeton, sought to gain both Einstein and Weyl for the new faculty. Weyl initially rejected the offer, but took up negotiations again after the dismissal of Courant and co.; see [Schappacher 1998, 524–525].

[6]Noether to Taussky, 24 April 1933, Papers of John Todd and Olga Taussky-Todd, Box 11, Folder 11, Caltech Archives.

One week before Noether wrote this letter, the Reichstag had gone down in flames as a result of arson. The day afterward, Hitler induced President Hindenburg to sign the Reichstag Fire Decree, which suspended most civil liberties in Germany; they would never again be restored under Nazi rule. New elections were called for March 5, the very day Noether wrote her letter. Having successfully suppressed the Communists (who formerly held 17% of the seats), the Nazis won an impressive victory, gaining 44% of the seats, enough to have a majority in alliance with the German Nationalist Party. On March 23, the newly assembled legislature convened in the Kroll Opera House, where it passed the Enabling Act that gave Hitler quasi-dictatorial powers. That event marked the end of the ill-fated Weimar Republic.

On March 30, still in Arosa, Courant wrote a remarkable letter to his colleague, the physicist James Franck. During this vacation, he, Rellich, and Kurt Friedrichs were hard at work on the long-awaited second volume of Courant-Hilbert *Methoden der mathematischen Physik*. Yet Courant had recently learned through his house servants that rumors were spreading about how he and his family would not be returning to Göttingen, so he sought Franck's advice. He had read that a nationwide boycott of Jewish businesses had been announced for April 1, and he hoped it could be deterred. He also expressed his anger over reports in the newspapers, citing public comments Einstein had issued about the political situation in Germany. Courant was appalled by this, since it made "Germany's internal situation a butt for general political agitation abroad" [Reid 1976, 139]. He also resented that Einstein had taken steps to renounce his German citizenship (he was still a Swiss citizen).

Shortly before his death, when Constance Reid showed him this letter and others, he could only comment that he really could not believe what he had written back then. Surely that applied to this passage from his letter to Franck:

> What hurts me particularly is that the renewed wave of antisemitism is ... directed indiscriminately against every person of Jewish ancestry, no matter how truly German he may feel within himself, no matter how he and his family have bled during the war and how much he has contributed to the general community. I can't believe that such injustice can prevail much longer – in particular, since it depends so much on the leaders, especially Hitler, whose last speech made a quite positive impression on me. [Reid 1976, 140]

On Franck's advice, Courant left his family behind in Switzerland and returned to Göttingen. He then learned that the SA had not only carried out the planned boycott on April 1st, they had five days earlier given the local Jewish community a foretaste of what was later to come by staging a march through the town, during which they demolished store fronts, vandalized property, and brutalized Jewish citizens. Several of those who committed acts of violence on that day were brought to court, though in all but one case the charges were dropped [Bruns-Wüstefeld 1997, 59–63]. No doubt many who participated – a group of well over 100 SA men – were among the perpetrators who burned down the city's synagogue during the infamous nationwide pogrom on November 9/10, 1938, the Night of Broken Glass.

On April 7, 1933, the government passed a new Law for the Restoration of the Professional Civil Service (BBG) [Schappacher 1998, 527–532]. The word "restoration" (*Wiederherstellung*) was, of course, a euphemism for purging undesirable elements from government employment. It contained an array of criteria, racial as well as political, and was directed not only against Jews but also against anyone whose political views might be regarded as suspect, which included those who were merely critical of the then fervent German nationalist ideology. Others, like Courant and Franck, who had fought in the war were to be exempted. Few then realized that this would only be the first among several waves of measures directed at individuals whom the Nazis portrayed as enemies of the German people.

Courant conferred with the physicists James Franck and Max Born as well as his trusted senior assistant, Otto Neugebauer.[7] The latter was Austrian and, unlike the other three, unaffected by §3 of the BBG, since he was an "Aryan."[8] They considered filing a symbolic protest, but Franck was considering a far more drastic step, namely to resign his professorship. The Prussian Ministry had extended the vacation period in order to implement the new law. Before the summer semester began on May 1, faculty members were to fill out questionnaires as part of a bureaucratic procedure to determine whether their personal biography met the conditions for dismissal.[9] Yet already by mid-April various individuals learned that they had been placed on leave of absence while their cases were under review.

James Franck, a Nobel laureate and highly esteemed member of the faculty, deliberated for a week before reaching his decision. On Easter Sunday, April 16, he wrote the Minister of Education announcing his resignation. He also wrote to the then serving Rektor, Siegmund Schermer,[10] and contacted the *Göttinger Tageblatt* by telephone [Rosenow 1998, 555–556]. The newspaper ran a major article that appeared on the day Franck's announcement reached the Ministry. It contained a statement from Franck that created a furor within the faculty:

---

[7]On Neugebauer's early career and his relationship with Courant, see [Rowe 2016].

[8]The term Aryan as used in §3 essentially meant non-Jewish, which would be defined more precisely in 1935 in the Nuremberg Laws.

[9]Statistics on the implementation of the BBG and other measures at 15 German universities were published in [Grüttner/Kinas 2007]; this study included adjunct lecturers, like Emmy Noether, who did not hold civil service positions.

[10]The Rektor (or Präsident) is the elected highest official at German universities. In Göttingen, a new Rektor, the Germanist Friedrich Neumann, was elected on May 1. Neumann joined the NSDAP that same day, and pursued policies fully in accord with the Nazi regime. On May 10, 1933, he opened the book-burning ceremony in Göttingen, part of a nationwide academic demonstration of extreme intolerance for literature deemed anti-German. After the war, Neumann was eventually rehabilitated and in 1971 Marburg University awarded him its Brother Grimm Prize.

We Germans of Jewish descent are being treated as aliens and enemies of our homeland. It is required that our children will grow up with the knowledge that they will never be able to prove themselves as Germans .... Those who fought in the war are supposed to have permission to continue to serve the State. I refuse to avail myself of this privilege, even though I understand the position of those who consider it their duty to remain at their posts.[11]

On April 24, the *Göttinger Tageblatt* published an article – signed by 42 professors (who claimed to be writing on behalf of all their colleagues) – that vehemently protested against the manner in which Franck had tendered his resignation, but especially the opening sentence cited above. They considered this not only an impediment to the government's domestic and foreign policies of national renewal but as "an act of sabotage," and expressed the hope that "the government would carry out the necessary purges expeditiously" [Hentschel 1996, 33]. Rektor Schermer had initially counseled delay with regard to Franck's case, but in view of the local headlines it was making he reversed course and requested that the Ministry respond quickly, since "the impression has grown that Franck was acting in concert with a specific group of faculty members" [Rosenow 1998, 557].

8.2 First Wave of Dismissals

These were the immediate background events that now prompted the Ministry of Education to take action.[12] On April 25, 1933, officials there sent a telegram to the University of Göttingen stating that six of its faculty members were to be placed on leave of absence, effective immediately. Three were mathematicians: Felix Bernstein, Richard Courant, and Emmy Noether, none of whom should have been dismissed according to the conditions stipulated in §3 of the BBG. Bernstein was already appointed associate professor in Göttingen in 1911, which meant he qualified for the exemption granted those who were already civil servants before the Great War. Courant served on the Western front during the war until September 1915, when he narrowly escaped death, and then returned as part of a unit engaged in underground telegraphy. Emmy Noether only held the honorary title of extraordinary professor, so she enjoyed none of the privileges of a civil servant. The fact that the government simply ignored this fact makes evident that the BBG was merely a pretext for removing those it deemed undesirable.[13] She had been forced to wait four long years before gaining the right to teach, the *venia legendi*, a title traditionally conferred by a university faculty and only routinely confirmed by the state. Now that the much-maligned Weimar Republic was dead and the "national renewal" had begun, the lion's share of the Göttingen

---

[11] Translated from [Rosenow 1998, 556].

[12] For an overview of the situation in Göttingen, see [Dahms 2008].

[13] In a third version of the original law, instated on 6 May 1933, it was extended to include employees in state positions who were not civil servants.

faculty clearly thought it high time for the state to take bold action; indeed, what better time to conduct a purge?

As it happened, the framers of the BBG had not neglected to include an elastic formulation for §4 that could serve as a catchall condition for removing those suspected of harboring sympathy for liberal causes. This read: "Officials, who by virtue of their previous political activities do not offer a guarantee that they will stand firmly for the national state at all times, can be dismissed from their positions." In this connection, the German League of Human Rights (DLfM) was just one of the organizations that the Nazis considered subversive.[14]

Courant was totally shocked when he learned about the Ministry's action, which was reported in the papers almost immediately. Having received no official information at all, he began to wonder whether this was merely an oversight, but he also worried that his brief activity as a local Social Democratic politician after the war might have been used to defame him. In a letter to his former assistant, Hellmuth Kneser, he mulled over all the possible reasons why he had been targeted, one being that unfriendly colleagues called his institute a "fortress of Marxism" [Reid 1976, 144]. Emmy Noether's case presented a similar problem, and she made no attempt to hide the fact that she, too, had taken an active interest in leftist politics during the early years of the Weimar Republic. In filling out the required questionnaire, she listed her membership in the pacifist-oriented Independent Social Democratic Party (USPD) from 1919 to 1922, after which she (like most USPD members) joined the mainstream SPD. After 1924, she dropped her membership altogether, and in a letter to Hasse she remarked, no doubt with a twinkle in her eye: "I never voted further left than that!"[15] She surely must have known that in the eyes of her conservative colleagues (and presumably Hasse's too), she was if not a Marxist, then at least a sympathizer. Many would have known that she spent a semester teaching in the Soviet Union, and that she and Courant spent ample time hosting Russian visitors. Although the axe had not yet fallen, the situation looked ominous for both of them. Yet Emmy just kept doing mathematics as before; she mainly seemed upset over the fact that her course on hypercomplex methods in number theory could no longer take place (though she later decided to convene with a small group of students in her new apartment). Courant, on the other hand, saw his whole life's work crumbling before his eyes. He was deeply troubled, but also determined to salvage whatever he could in this bizarre situation.

Courant was obviously a high-profile figure, and so his case was dealt with differently than Emmy Noether's. On legal grounds, he could only be dismissed in accord with §4, which meant determining that he was politically suspect. Weyl obtained a detailed report from Courant about his past military and political

---

[14]The DLfM was explicitly named on the Ministry's questionnaire; Otto Blumenthal's membership in this organization led automatically to his dismissal from his professorship in Aachen [Rowe/Felsch 2019, 360–374].

[15]Noether to Hasse, 21 July 1933, [Lemmermeyer/Roquette 2006, 197]; Noether's politics are discussed in [McLarty 2005].

activities, which he forwarded on 23 May to the Kurator,[16] Theodor Valentiner, who sympathized with Courant. In doing so, Weyl emphasized that Courant's report was "of the greatest importance for judging the case . . . and must not be overlooked" [Schappacher 2000, 21]. Valentiner, a jurist who had previously served as Kurator from 1921 to 1932, attached a lengthy personal and legal analysis to these documents. Concerning Courant's "political sins" during the early Weimar years, he wrote:

> In any case, I have clearly sensed, as I have told trusted people as early as 1925 or 1926, that he evidently had found his way back to the middle class. . . . my confidential inquiries with suitable, thoroughly right-wing professors led to the result that no one remembers a single utterance from his mouth since 1919 that could justify a different view.
> . . .
> There is no doubt in my mind that he will stand up for the national state at all times, and he made this declaration directly to me. [Schappacher 2000, 22]

While Courant's case dragged on, Nazi students were already pushing forward their preferred candidate for his chair: the Darmstadt mathematician Udo Wegner. As the end of the semester neared, Weyl wrote a lengthy memorandum on the sad state of mathematics in Göttingen, underscoring the need to hire suitably qualified personnel, and pointing in particular to the potential vacancy that might arise with Courant's professorship. By now he surely realized that Courant's position was untenable, so he took the opportunity to attack the politically motivated candidacy of Wegner, which he predicted would be "catastrophic for mathematics in Göttingen" [Schappacher 2000, 23]. Soon thereafter, Weyl left to vacation in Switzerland. From there, he renewed negotiations with Flexner, who was fully informed about the situation at the German universities. In October 1933, Weyl informed the Ministry that he had accepted an offer from the Institute for Advanced Study.[17] In the meantime, Courant's case remained undecided, but he arranged to take an official leave of absence (without pay) in order to spend the coming academic year in Cambridge. Leaving his family behind, he began to contemplate new plans for a future life in the United States.[18]

Before turning to Noether's case, it might here be noted that she and Courant were two among the 52 members of the Göttingen faculty who lost their positions directly as a result of Nazi policies. To put that number in perspective, this represents slightly more than one-fifth of the entire teaching corps; Noether was the only woman in this group (her dismissal then left one other woman on the fac-

---

[16]The Kurator was the official representative of the Minister at German universities.

[17]He did not neglect to add that, since his wife came from a Jewish family, he presumed the Ministry would have no objection to releasing him from his position in Göttingen [Schappacher 2000, 24].

[18]His case was eventually dropped after Courant negotiated reasonable conditions for his family to emigrate; in 1934 he began a new career at New York University [Reid 1976, 155–168].

ulty).[19] Forty of these cases were due to racial discrimination, and in 32 instances the dismissals led to emigration. Three individuals voluntarily resigned (Franck and Born being two such cases), raising the total losses to 55. The numbers (both absolute and in percentage) were even higher at four other German universities: Frankfurt, Berlin, Heidelberg, and Hamburg [Grüttner/Kinas 2007, 140,166].

In view of this sudden and sweeping calamity, it should come as no surprise that very few of these dismissals drew efforts even remotely comparable to those undertaken on behalf of Courant. Still, the testimonials in support of Emmy Noether were every bit as strong, perhaps even more so, thanks to the efforts of Helmut Hasse, who intervened just as the meltdown in Göttingen had begun. Very few of his letters to Noether have survived, but she answered one of those many lost letters on the 10th of May:

> Thank you very much for your most friendly letter! Still, the matter is much less unfortunate for me than for many others: first, I have a small savings (I was never entitled to a pension); at the moment, I'm still collecting a salary, so I can wait until the definitive decision or a little longer. In the meantime the faculty will try what it can to prevent termination; the success of that, though, is quite doubtful at the moment. Finally, Weyl (see Fig. 8.1) told me that a few weeks ago, when everything was still up in the air,[20] he had written to Princeton, where he still has contacts. ...Weyl thinks that something could open up over time, especially since last year Veblen was keen to let Flexner know about me.[21] ...This "do not teach until further notice" is quite catastrophic here at the institute ...I'm reading your lectures with great pleasure; I think that once in a while I'll invite the "Noether community" (Noethergemeinschaft)[22] to my apartment to talk about them. [Lemmermeyer/Roquette 2006, 187–188].

8.3 Hasse's Campaign for Noether

Hasse contacted Neugebauer before proceeding with his plan, namely to approach "a number of well-known mathematical scholars from Germany and abroad for assessments of the scientific significance of Miss Noether's work and her entire scientific personality."[23] Beyond his personal relationship with Emmy Noether,

---

[19]Women held slightly more than 1% of the teaching positions at the German universities; more than one-third of them lost their positions due to NS-policies [Grüttner/Kinas 2007, 142].

[20]Presumably this means before the Ministry's telegram from 25 April had arrived.

[21]It appears likely that Veblen introduced Emmy Noether to Flexner during the International Congress held in Zurich the previous September. [Lemmermeyer/Roquette 2006, 189].

[22]This is a play on the word Notgemeinschaft, reflecting her own situation and the state of science in general, since the Notgemeinschaft der deutschen Wissenschaft, founded in 1920 and the forerunner of the present-day German Research Foundation, was established expressly as an Emergency Foundation.

[23]Hasse to Kurator of Göttingen University, 3 June 1933, translated from [Roquette 2008].

Figure 8.1: Emil Artin in Göttingen, July 1933: l. to r. Ernst Witt, Paul Bernays, Helene, Hermann, and Joachim Weyl, Artin, Emmy Noether, Ernst Knauf, unidentified person, Chiungtze Tsen, Erna Bannow. Photo by Natascha Artin; for dating and details, see [Eckes/Schappacher 2016] (Auguste Dick Papers, 12-14, Austrian Academy of Sciences, Vienna)

Hasse had a second motive for supporting her case. In his letter to Valentiner, the Kurator, he expressed this in these words:

> I also intend to draft a detailed report of this kind on my own, since I am convinced that Miss Noether is one of the leading German mathematicians and that German science, and especially the younger generation, will suffer a very serious loss should Miss Noether be forced indirectly to move abroad. As I hear, Miss Noether's special students have also prepared a report that should be sent to you in a short time.

Indeed, Wolfgang Wichmann soon submitted this testimonial, undersigned by friends of Emmy Noether as well as her students in the narrower sense.[24] She was certainly well aware of this text, which to a remarkable degree sought to paint

---

[24]The signatories were E. Bannow, E. Knauf, Tsen, W. Vorbeck, G. Dechamps, W. Wichmann, H. Davenport (Cambridge, Engl.), H. Ulm, L. Schwarz, Walter Brandt (?), D. Derry, and Wei-Liang Chow; see [Roquette 2008].

her as a kind of nativist German mathematician (*eine urdeutsche Mathematikerin*). Carl Ludwig Siegel's report pointed in a similar direction, though in a much subtler manner:

> Through investigations by Dedekind, Frobenius, and Kronecker, Germany attained a leading position in algebra toward the end of the last century. That the country has maintained and even strengthened its position until this day is largely due to Emmy Noether. In particular, Miss Noether's publications and lectures have greatly promoted the so-called theory of hypercomplex systems so that the problems arising from it are now the focus of interest among algebraists around the world.
>
> In the past decade, E. Noether has stimulated productive work among young mathematicians probably more than any other Göttingen docent. In the interest of our young academics, it is therefore urgently wished that Miss Noether be able to continue teaching in Germany.[25]

The latter group went a good deal further in setting forth their case that Noether belonged in Germany:

> We doctoral students of Prof E. Noether, students of mathematics at the local university, request that the following considerations be taken into account: As much as we welcome the national revolution and all its effects, we also regret that Prof. Noether was placed on leave of absence, thereby preventing her from carrying out her work effectively, and this for the following reason. Ms. Noether has founded a mathematical school that has brought forth the most capable of the younger mathematicians, some of whom are now lecturers or professors at German universities. Her work has always consisted of special lectures with small groups of auditors, most of whom wish to pursue an academic career. The fact that her courses span over several semesters has also meant that students are given a deeper insight into the interconnections.
>
> It is no coincidence that all her students are Aryans; this is due to her essential conception of mathematics, which corresponds entirely to an Aryan way of thinking. This does not concern detached individual results, but rather a way of recognizing and understanding the whole, and E. Noether succeeds in doing this based on a conceptual method that she has developed in recent years. Through the research field she explores and the lively questions she poses, she has filled all of her students with enthusiasm and passion for mathematics.
>
> Despite differing political views, our personal relations with her are in no way disturbed, as she has never had any political influence on her students. The close connections she has managed to establish

---

[25]C.L. Siegel, Frankfurt (Main), 14 June 1933, [Roquette 2008].

between herself and her students, as well as among the students themselves, comes from her great personal stimulation. These connections can hardly be maintained for long without further contact. Some students have already left this semester for other universities. This is the reason why we would be pleased if Professor Noether were given the opportunity to exercise again her profession as a teacher, one who is unique in all of Germany.

Emmy Noether was not only well aware of this ongoing effort on her behalf, she took an active part in planning it. In particular, she surely read and presumably approved the above text with its assertion that Noether's mathematics "corresponds entirely to an Aryan way of thinking." Could this have been intended as a ploy to counter §3 in the BBG, which clearly applied to her as a non-Aryan?[26] One also has to wonder why this text brought up the political differences between Noether and her "good Aryan students." In any case, she did take an avid interest in mobilizing support, both from within Göttingen as well as from the mathematical world at large, as can be seen from her letter to Hasse, dated 21 June 1933:

> You really are making quite a job for yourself with the reports! As if you didn't otherwise have enough work to do! Wichmann had just given the Kurator the report with student signatures – mainly from the algebraists – before he [Valentiner] left for Berlin, which the latter said was very correct, though of course at the moment it is difficult to get past §3. Then he got your letter as well! It would, however, seem to me good, in case a sufficient number of reports have arrived by the end of the semester, to send these already to the Kurator and forward the others (Takagi etc.) later; it has been said that the matter will not be decided beforehand – i.e., during the semester. And it also seems to me a good idea to make copies of the reports beforehand (but at my expense!), so that it will be easier to refer back to them later should they not be successful this time. [Lemmermeyer/Roquette 2006, 189–190]

Emmy spent early August in the small beach resort town of Dierhagen on the Baltic, where she met her brother Fritz and his family (Figs. 8.2, 8.3). They were joined by Herbert Heisig and his wife Lotte, together with Hans and Eva Baerwald. Heisig was a native of Breslau and, like Baerwald, studied under Noether at the Institute of Technology; in 1931 he took his doctorate in engineering mathematics there. Only a few months before they vacationed together, Herbert Heisig was appointed head of the mathematics and natural sciences division at the Teubner publishing firm in Leipzig.[27] Fritz Noether was in nearly the same situation

---

[26]At the time Noether was placed on leave, the BBG applied only to those who were civil servants, but soon afterward it was extended to anyone certified to teach at an institution of higher education.

[27]Heisig remained with Teubner-Leipzig until 1952 when he assumed the same position for the West German branch in Stuttgart, which he headed until 1969.

Figure 8.2: Emmy with Fritz and Regina Noether and Herbert and Lotte Heisig, Dierhagen, August 1933 (Auguste Dick Papers, 12-14, Austrian Academy of Sciences, Vienna)

as his sister, waiting to receive definitive news about whether he would lose his professorship at the Breslau Institute of Technology. He was hired there in 1922, one year after Emmy's protégé Werner Schmeidler assumed Max Dehn's chair in Breslau.

On 21 July, before she left for vacation, Noether wrote to Hasse. She was excited about his new results on a function-theoretic version of the Riemann hypothesis and asked if he would lecture about that at the forthcoming DMV meeting in Würzburg. Obviously, she wanted very badly to attend that meeting, but felt she needed encouragement to do so under the circumstances. She had in the meantime spoken with Weyl, who told her he would find out from the Kurator about the submission date for the reports Hasse had gathered. Weyl assumed that would be soon, but he also agreed that copies ought to be made in advance, whether in Marburg or in Göttingen was a matter of indifference to her. Hasse had probably inquired as to whether a de facto case could be made that she would have qualified as a private lecturer before the outbreak of the war – a potential argument for gaining an exemption to §3. In the meantime, she had filled out the questionnaire for the Ministry, and so informed Hasse: "I stated that Klein and Hilbert brought me to Göttingen in the spring of 1915 to fill in for the private lecturers. In order to conclude from this that I already met all the preconditions in August 1914 would require quite a lot of imaginary benevolence!" [Lemmermeyer/Roquette 2006, 196].

On July 31, the day Noether left for the Baltic, Hasse submitted all 13 reports to Valentiner, the Göttingen Kurator. Hermann Weyl may have received

Figure 8.3: From left to right: Hermann Noether, his girlfriend Nora, Emmy Noether, Eva Baerwald, Fritz Noether, Lotte Heisig, Regina Noether, Herbert Heisig, with Hans Baerwald in the foreground, Dierhagen, August 1933 (Archives of the Mathematisches Forschungsinstitut Oberwolfach)

them first or else read them shortly afterward. In his memorial lecture for Emmy Noether at Bryn Mawr on 26 April 1935, he remarked: "I suppose there could hardly have been in any other case such a pile of enthusiastic testimonials filed with the Ministerium as was sent in on her behalf. At that time we really fought; there was still hope left that the worst could be warded off. It was in vain" [Weyl 1935, 435]. In his own testimonial, Weyl compared her with the two most famous women in the recent history of mathematics, Sophie Germain and Sofia Kovalevskaya, claiming that Noether surpassed them both in her originality and depth. He also took up the defense of "abstract algebra" as practiced by Noether, who grasps problems with "seeing thoughts and through the formation of concepts as appropriate as possible for the object, rather than through blind calculation. In this respect Ms. Noether is the legitimate successor of the great German number theorist R. Dedekind." Moreover, thanks to quantum theory, algebra is the area of mathematics that stands in the most intimate relationship with physics, and "in this field, in which mathematical research is currently developing most rapidly,

Emmy Noether is recognized at home and abroad as the true leader" (Hermann Weyl, Göttingen, 12 July 1933, [Roquette 2008]).[28]

Hasse clearly took some care in selecting the foreigners whom he asked to write on behalf of Noether, drawing in part on his own personal connections in the world of algebraic number theory. He evidently also informed her about whom he had contacted, since she referred directly to Takagi as one of these persons. Had he asked for her advice directly, she surely would have suggested Kenjiro Shoda, but Takagi reached this conclusion himself, so there were two letters from Japanese mathematicians. Shoda expressed his personal appreciation for his former teacher with these words:

> Frl. Noether, who has brought forth so many new and fundamental ideas in the theory of hypercomplex numbers, representation theory, ideal theory, etc., is generally regarded by us foreigners as the most outstanding representative of German algebra. Like so many other Japanese algebraists, I remember with special thanks the time in Göttingen when I studied with Ms. Noether and gained so much invaluable scientific and personal encouragement from her. All of us wish very much that your efforts shall succeed in maintaining Ms. Noether for German mathematics. (Kenjiro Shoda, Osaka, 16 July 1933, [Roquette 2008])

Noether may have suggested the name Beniamino Segre, who was Severi's assistant in Rome in 1928, the year she and Segre likely met at the Bologna ICM. Three years later, Segre gained a professorship in Bologna, but as a Jew he fell victim to the racial laws implemented in 1938, which forced his emigration to England. His testimonial for Emmy reflects the reverence Italian geometers held for her father.

> The tremendous scientific value of Max Noether's geometric work is recognized by everyone who has been profoundly influenced by Riemann's immortal work on the theory of algebraic functions and their integrals. This formed the starting point for an astonishing flowering of studies in France and especially in Italy since 1890, so that Max Noether can rightly be regarded as the founder of the great structure of today's algebraic geometry.
>
> Ms. Emmy Noether is the worthy successor of the paternal name, although her works have a somewhat different direction and, above all, a purely algebraic emphasis. Miss Noether's work is both remarkable and important. Together with Artin, Hasse, and van der Waerden, she is the recognized head of a school in modern studies of abstract

---

[28] In connection with quantum physics, Weyl was alluding to the role of group representation theory, which had recently been elaborated in monographs by Eugene Wigner, van der Waerden, and Weyl himself. Noether's more abstract approach to this part of modern algebra was largely independent of these currents, although Martina Schneider has noted how [van der Waerden 1932] drew on the theory of groups with operators, a topic developed by Wolfgang Krull and Emmy Noether [Schneider 2011, 191–205].

algebra and general number theory that continues and crowns the fundamental ideas of Grassmann, Dedekind, Kronecker, and Weierstrass. (Beniamino Segre, Bologna, [Roquette 2008])

Two of the testimonials came from Viennese number theorists, Philipp Furtwängler and Anton Rella, neither of whom knew Noether well. Furtwängler offered this opinion: "Ms. E. Noether holds a leading position in the development of modern algebra. Not only has she furthered this theory partly through her own work and partly through the work of others, she has also through her selfless teaching, led by ideal goals alone, established a large circle of students, who have already today made a name for themselves in the mathematical world" (Philipp Furtwängler, Vienna, 29 June 1933, [Roquette 2008]). Rella was Gottfried Köthe's doctoral adviser, which suggests perhaps a somewhat closer relationship to Noether's sphere of influence. Rella apparently did meet her at the congresses in Bologna (1928) and Zurich (1932), but otherwise refers to secondhand knowledge of her enormous talents as a teacher:

> It actually seems inappropriate for me to offer a statement about a scientific personality of such eminence as Emmy Noether.
> I regard Ms. Noether as simply the leading figure in her special field of abstract algebra. From reports of my own former students who continued their mathematical training in Göttingen, I know that Ms. Noether, in her impulsive manner, is able to exert the greatest scientific influence through personal communication with young mathematicians, as is shown by the large number of leading young algebraists in Germany, almost all of whom regard themselves as her personal pupil or at least have been inspired by her to carry out their own research. (Anton Rella, Vienna, 9 July 1933, [Roquette 2008])

As one of Richard Courant's closest collaborators, Harald Bohr was a frequent visitor in Göttingen. His interests in analytic number theory also drew him to Göttingen's Edmund Landau, whom many considered the leading authority in this field. Another leading number theorist was the Cambridge mathematician G.H. Hardy, who happened to be visiting Bohr in Copenhagen when the latter received Hasse's request. The testimonial Bohr wrote was signed by both men:

> Miss Noether is of paramount importance for the development of modern algebra and she is rightly regarded as the head of a school of young algebraists in and outside of Germany. The fact that algebra has experienced a new flowering and now stands at the forefront in the entire mathematical world, expanding into geometry and other areas of research, this has mainly been due to Miss Noether and her school. The influence this has had reaches far beyond the borders of Germany, and throughout the world hers is one of the best known names. (Harald Bohr and G.H. Hardy, Copenhagen, August 1933, [Roquette 2008])

The Swiss mathematician Andreas Speiser was another number theorist who took his doctorate in 1911 under Hilbert in Göttingen, although he was actually a student of Hermann Minkowski. Speiser also had strong interests in modern algebra. He arranged that Johann Jakob Burckhardt, his assistant in Zurich, translate Dickson's seminal algebra book (*Algebren und ihre Zahlentheorie*, [Dickson 1927]), for which Speiser added an appendix on ideal theory. He answered Hasse's request by writing in the name of all mathematicians in Zurich:

> [Emmy Noether] is undoubtedly the most important living female mathematician and possesses a highly extensive knowledge. We invited her to deliver a plenary lecture at the international congress in Zurich in 1932 because she is one of the leading experts on modern algebra and especially since she works at the center of an excellent school in Germany. She brought the important ideas of Wedderburn and others to fruition, and as an extremely stimulating personality and creative mathematician she has exerted the greatest influence in her country. Since she lives for her science like few others, she also exercises a moral influence that should not be overlooked. To us she appears to be indispensable for Göttingen's reputation and influence in the mathematical world. All of the younger mathematicians who studied there went through her school. (Andreas Speiser, Zurich, 1 July 1933, [Roquette 2008])

Two testimonials were written by Dutch mathematicians, one by B.L. van der Waerden. As one of her closest associates, he was especially well qualified to judge Noether's work and its significance.

> Dr. Emmy Noether is a personality of unique importance in the mathematical world. Some 13 years ago, she began indicating the direction in which algebra and arithmetic should develop in her opinion, and now, in fact, they are developing under her recognized leadership in this very direction. She has held on to her own methods and problems with firm energy, even in times when the problems were considered too abstract and the methods too sterile, and now these methods have been successfully applied everywhere, especially in Germany, but also in France, Holland, Russia, America, and Japan, and they have delivered the most beautiful results. Before her leave of absence, algebraists from all over the world came to Göttingen to learn her methods, get her advice, and to work under her leadership. Her merits were recognized by the Faculty of Mathematics and Natural Sciences in Leipzig and by the German Mathematical Society, which in 1932 awarded her and E. Artin the Ackermann-Teubner Memorial Prize. (B.L. van der Waerden, Leipzig, 8 June 1 933, [Roquette 2008])

One might have anticipated that the second Dutch mathematician to write on Noether's behalf would have been L.E.J. Brouwer, since he had known her even longer than had van der Waerden. Little had been heard from Brouwer, however, after he was dramatically purged from the editorial board of *Mathematische Annalen* in December 1928, and even Pavel Alexandrov was no longer in touch with him as before. Instead, Brouwer's old nemesis, Jan Arnoldus Schouten, wrote an impassioned report, in which he expressed his outrage over the Nazi racial policies. Given the delicacy of the situation, one wonders whether Hasse might not have had second thoughts about including such a forthright political statement among these testimonials.

> In my opinion, Miss Noether is a first-class mathematician, indeed, she is the greatest living female mathematician in the world! Through hard work and sacrifice, she has done an incredible amount for mathematics and for her numerous students. Many of her students live here in Holland and gratefully acknowledge how much they have learned from her. ... It would be a great scandal if such a powerful figure were lost due to racial prejudice. Apparently people in Germany have no conception of the outrage over this in those foreign countries friendly toward Germany. This is not a matter of Jews being beaten up here or there by misguided youth. What is happening here through the official authorities themselves causes the utmost anger among us, the friends of Germany. Only France and the Little Entente can feel happy!
>
> ...
>
> For orientation, I am of purely Aryan descent and 50% German, my mother comes from a family of Prussian officers; thus, in the eyes of the current rulers, of absolutely unobjectionable heritage. Before, during, and after the war, I was always extremely pro-German, which brought me a good deal of unpleasantness. I have never been ashamed of Germany and of my parentage, though I would be now if I were not of the conviction that what is happening is completely contrary to the essence of the German people's soul. (Jan Arnoldus Schouten, Delft, 27 June 1933, [Roquette 2008])

Emmy Noether and the Munich mathematician Oskar Perron would seem an unlikely pairing. Perron's tastes ran strongly toward classical mathematics, including number theory, with no signs of interest in abstract mathematics. Yet his report reflects real understanding of Noether's work and the nature of her influence, which makes it highly likely that he knew her personally.

> Emmy Noether belongs among the leading personalities in modern mathematics. Already her older work on invariant theory testifies to a high level of skill; her studies on the general theory of fields and ideals are groundbreaking. Yet even more than through her printed publications, Emmy Noether has always understood how to inspire interest in young people for great scientific ideas through oral communication and personal interchange, and so a whole younger generation of algebraists has been influenced by her to a high degree and follows in her footsteps.

One, however, who took inspiration from her spirit can no longer walk in her footsteps because he died in the war. For him, she erected a scientific monument with touching humility and piety; without Emmy, the name Kurt Hentzelt would be forgotten today. (Oskar Perron, Munich, [Roquette 2008]).

Figure 8.4: Emmy Noether departing from Göttingen for the USA, October 1933 (Courtesy of MFO, Oberwolfach Research Institute for Mathematics)

Helmut Hasse's somewhat lengthier testimonial also emphasized that Noether's mathematics was firmly rooted in a Germanic tradition. Like Weyl, but with a far stronger emphasis on her ties to Germany, Hasse sought to deflect the idea that Noether's abstract mathematics was somehow "alien" (artfremd) or without substance.

Miss Noether is the founder and leader of the modern algebraic school that developed in Germany after the war. Through a series of profound works that tie in with the life's work of the German mathematician Richard Dedekind, and through her personal influence on numerous young German mathematicians, she laid the foundation for the transformation of traditional algebra ... while continuing to do so through her own collaborations as well as the impulses she gives to her numerous students.

A flourishing school, comprised largely of young German mathematicians, would be robbed of its universally recognized leader if Miss Noether were forced to go abroad as a result of the current political situation. Most of these younger German mathematicians, those who have directly or indirectly gone through her school, engage with her in lively exchanges of ideas during frequent visits to Göttingen or at the conferences of German mathematicians, meetings that have a particularly stimulating and fruitful effect on their own work. All these mathematicians, and thus German mathematics in general, would suffer significant nonmaterial damage should the opportunity for such personal exchanges of ideas be taken away.

Miss Noether has always felt like a German, and she still has a strong desire to remain in Germany in her position at the Göttingen Mathematical Institute, where her entire personality belongs. There, like nowhere else in the world, she has the opportunity to meet the kinds of people needed to pursue her far-reaching ideas, which she can only work out to a small extent herself. She can then, over and again, point them toward the research paths she has in mind. Because there alone, following an old tradition, one finds a select stream of mathematically talented youth capable of taking up these ideas, at least in higher semesters.

In no sense can one call her mathematics "alien". On the contrary, it has a quality much like the typical German mindset, which in its nature favors the intellectual, the theoretical, and the ideal rather than such qualities as purpose, material success, or the real. That this is so can be seen from the fact that the vast majority of German mathematicians who have found their way to her school over the past two decades are of Aryan descent. As proof of her truly outstanding significance one may further point to the fact that she is recognized as a leading contemporary mathematician even in Anglo-American countries, which favor the completely opposite orientation with more reality-based mathematical conceptions.

Finally, it should be noted that Miss Noether's father, Max Noether, was an outstanding German mathematician, who in his day received an honorary award from the Academy in Rome for his achievements in algebraic-geometric fields. Still today, for the important new Italian school of mathematics, he enjoys the greatest esteem among all German mathematicians. (H. Hasse, Marburg, 31 July 1933, [Roquette 2008])

Since Emmy Noether had already left on her vacation when Hasse sent these reports to Göttingen, she could not have read the original texts. She must have read them later, however, since Hasse (or possibly Weyl) ordered that copies of them be made, as she had requested. Emmy clearly felt a great deal of gratitude for the support she received during such a difficult time, even knowing that a positive outcome was improbable. In his cover letter to Valentiner, Hasse again stressed the exceptional nature of the situation, but also its limited purpose, namely "to maintain her existence in any form at the Göttingen Mathematical Institute," to which he added:

> Not only for Göttingen, but for German mathematics in general, it would be an immeasurable loss if Miss Noether found no further opportunity in Germany to continue teaching mathematics. Since her teaching stands outside the framework of the training plan for teaching candidates, but instead involves the stimulation of a relatively small group of advanced students who mostly have academic careers in mind, I dare to hope that such an activity might not completely cross with the basic considerations and principles that led to her temporary leave of absence.[29]

Unlike in Courant's case, Valentiner took a very critical view of Emmy Noether's political views and past history, though it seems doubtful that he actually spoke with anyone who had real knowledge of these matters. In forwarding the reports that Hasse submitted to the Ministry, he added these comments:

> Although I am aware of Miss Noether's scientific importance, I refrain *in this regard* from commenting on the expert opinions herewith enclosed from a number of competent authorities. From a political point of view, to my knowledge Miss N. has stood on Marxist ground from the time of the Revolution of 1918 to the present day. And although I hold it for possible that her political views are more theoretical than conscious and practical, I do believe with certainty that her sympathies so strongly favor a Marxist political worldview that she cannot be expected to stand up wholeheartedly for the national state.[30] So with all due respect for the scientific importance of Miss Noether, I am unable to support her.[31]

Whether this remark played any role at all is impossible to know; it seems most likely that her case was handled in a routine manner based on the fact that she failed to satisfy §3, the Aryan paragraph.

Emmy Noether returned from her vacation around September 1 and wrote to Hasse a few days afterward, thanking him again for all his efforts. She then

---

[29] Translated from [Tollmien 1990, 206].

[30] The German reads "ein rückhaltloses Eintreten für den nationalen Staat", in conformance with §4 of the BBG.

[31] Göttingen Kurator to Prussian Ministry, 9 August 1933, translated from [Tollmien 1990, 206].

remarked: "If not for now, then the reports may help later! And it seems only right to me that they are now available!"[32] This statement thus seems to confirm that copies of the reports were indeed made. Emmy had in the meantime begun making plans for the coming academic year:

> You have heard from Davenport that I want to go to Oxford for a term after Christmas. In the meantime, I received another offer for a research professorship in Bryn Mawr for one year (1933/34), but which I accepted for the following 34/35. I don't have an answer yet, but I doubt that the postponement – I can't very well be in England and America at the same time – should cause any problems. It's a joint fellowship from Rockefeller and the Committee in Aid of Displaced German Scholars.[33] Bryn Mawr is, by the way, another woman's college, but as Veblen later wrote me, it's the best of these; it's also so close to Princeton that I can come over often.... [Lemmermeyer/Roquette 2006, 199]

Noether further reported meeting Hamburg's Wilhelm Blaschke and Hans Rademacher from Breslau during her vacation on the Baltic, where they discussed the forthcoming DMV conference in Würzburg. Blaschke counseled that the DMV should "maintain its purely scientific, neutral character" and not take up the politically sensitive matter of the government's recent actions.[34] Emmy agreed with this, and thus wrote Hasse that she would probably not be coming to the Würzburg conference, one of the very few annual meetings of the DMV she did not attend.

One week later, on September 13, Noether wrote Hasse again to report that she had now received definitive news: her teaching certification had been withdrawn on the basis of §3 of the BBG, thus owing to her racial status. Payment of her salary would then terminate at the end of the month, leaving her no choice but to go abroad (Fig. 8.4). She thanked Hasse again for his efforts, and reassured him that the reports could well be valuable later. Max Deuring, her star student, would be coming to the DMV meeting in Würzburg, so Hasse would be able to learn about his and her latest ideas from Deuring.

As will be described in the next chapter, Emmy's conjecture that she could postpone accepting the offer from Bryn Mawr was incorrect. Her plans then quickly shifted and she readied herself for the transatlantic voyage that brought her to the small college outside Philadelphia that would become her new academic home. The following section offers a brief account of mathematics at Bryn Mawr College in the years preceding Noether's arrival.

---

[32]Noether to Hasse, 6/7 September 1933, [Lemmermeyer/Roquette 2006, 199]. The only known copies, however, are those in the *Geheimes Staatsarchiv Preußischer Kulturbesitz*, Berlin, available online at [Roquette 2008].

[33]The Emergency Committee in Aid of Displaced Foreign Scholars was created in 1933 to assist those who were barred from teaching by the Nazis. The program was later expanded to include Austria, Czechoslovakia, Norway, Belgium, the Netherlands, France, and Italy.

[34]Rademacher felt, on the contrary, that those who had been placed on leave should feel free to attend; in February 1934 Hans Rademacher lost his professorship in Breslau under §4 of the BBG (see Section 9.3).

Chapter 9

Emmy Noether in Bryn Mawr

9.1 Bryn Mawr College and Algebra in the United States

In the annals of higher education for women, two elite colleges were particularly important for mathematics: Girton College, in Cambridge, England and Bryn Mawr College, near Philadelphia, Pennsylvania. Both had significant historical ties with Göttingen.[1] A central figure in this story was Charlotte Angas Scott, who studied at Girton from 1876–1880. After completing her studies, she finished eighth in the Tripos examination, though her achievement went officially unacknowledged. Later, in 1885, she was awarded a doctorate from the University of London on the basis of an external examination. That same year, Scott joined the founding faculty at Bryn Mawr (Fig. 9.1), where she taught until her retirement in 1924. Two of her seven doctoral students, Isabel Maddison and Marguerite Lehr, also became fixtures of the Bryn Mawr faculty [Green/La Duke 2009, Green/La Duke 2016].[2]

Maddison had studied alongside Grace Chisholm at Girton College. Then, in the mid-1890s, both went on to do graduate work under Klein in Göttingen, the first German university to spearhead opportunities for (foreign) women to earn doctoral degrees. By this time, females had already gained the right to study at Paris University, though none had yet taken an advanced degree in mathematics. Scott's first doctoral student, Ruth Gentry, spent one semester studying at the University of Paris. Chisholm took her Ph.D. magna cum laude in 1895 after passing the usual qualifying examinations. Sofia Kovalevskaya's case twenty years earlier was altogether different; she never set foot in Göttingen and earned her doctorate in absentia, a quite common practice that the Prussian Ministry prohibited soon thereafter. In their study of American women mathematicians during the period 1891 to 1906, Fenster and Parshall found that of the 18 most active

---

[1] On Girton, see [McMurran/Tattersall 2017]; on Bryn Mawr, see [Parshall 2015].
[2] Three of her students were recently portrayed in [Lorenat 2020].

© The Author(s), under exclusive license to Springer Nature Switzerland AG 2020
D. E. Rowe, M. Koreuber, *Proving It Her Way*, https://doi.org/10.1007/978-3-030-62811-6_9

fully half had gone to Göttingen for part of their studies [Fenster/Parshall 1994, 241].

Figure 9.1: Faculty and Students at Bryn Mawr College, Spring 1886: standing just left of the entrance is Charlotte Angas Scott, seated next to her is M. Cary Thomas, first dean and second president of the college, seated in the middle is President James E. Rhoads, and standing at the right of the entrance is Woodrow Wilson, who taught history and political science (Bryn Mawr College Special Collections)

As the only college for women in the United States with a doctoral program in mathematics, Bryn Mawr naturally drew on talented graduates from various undergraduate institutions around the country, but especially the Seven Sisters colleges, to which Bryn Mawr belonged. The others were Mount Holyoke, Smith, Wellesley, and Vassar, all four independent liberal arts colleges for women, as well as Radcliffe and Barnard, which were associated with Harvard and Columbia, respectively. Hunter College in New York City, which specialized in training teachers, was another important outpost for aspiring female mathematicians. In their study of women who took doctoral degrees in the United States during the first four decades of the twentieth century, Judy Green and Jeanne LaDuke found that females constituted over 14 percent of the degrees granted, a proportion that would not be reached again until the 1980s [Green/La Duke 2009].

In 1896, Bryn Mawr established a Mathematical Journal Club, which met every other week to hear special reports and lectures presented by faculty members as well as graduate students. Two distinguished professors from nearby Haverford College, Frank Morley and E.W. Brown, also gave talks at these meetings. Bryn Mawr also offered a European Fellowship Program to outstanding students, several of whom took up studies in Göttingen. Little wonder that Felix Klein took a special interest in Bryn Mawr College, where his youngest daughter, Elisabeth, spent a year abroad before completing her studies in 1911. Her father briefly visited the college in 1896, when he met with Charlotte Angas Scott. She wrote him the following year to report: "I am expecting to send two of my best students to Göttingen next year; to both of them have been awarded College Fellowships . . . . One of them you met when you were here that Sunday afternoon."[3] Scott was referring to Virginia Ragsdale and Emilie Norton Martin, who spent the academic year 1897/98 studying with Klein and Hilbert. They were joined by a third Bryn Mawr student, Fanny Gates, along with Anne Bosworth, a graduate from Wellesley College who went on to take her Ph.D. under Hilbert. Martin afterward returned to Bryn Mawr and took her doctorate there in 1901. She later joined the faculty at Mount Holyoke College, where she went through the ranks from instructor to full professor, serving as department chair from 1927 through 1935.

By the time Emmy Noether habilitated in Göttingen in 1919, far fewer Americans were studying there. Still, she clearly made an effort to stay abreast of mathematical research in other countries, including the United States.[4] Noether reviewed Dickson's preliminary article [Dickson 1923a], which preceded the publication of his book *Algebras and their Arithmetics* [Dickson 1923b]. Dickson's book was also reviewed by Hazlett in [Hazlett 1924], where she pointed out that its approach drew on the earlier work of Wedderburn on division algebras. Four years later, Dickson's book appeared in a revised German edition, [Dickson 1927], which received a lengthy review in [Hasse 1928]. These reviews in some ways reflect the oft-mentioned tension between American and German algebraists during this era.[5]

Olive Hazlett was Dickson's second female doctoral student, following Mildred Sanderson, who died of tuberculosis at age 25. In a tribute to the latter,

---

[3]Scott to Klein, 19 March 1897, cited in [Green/La Duke 2016, Entry: Martin].

[4]This is confirmed by numerous reviews she wrote for the *Jahrbuch über die Fortschritte der Mathematik*, long the premier abstracting journal for mathematics. Among the American authors whose works she abstracted were Leonard Eugene Dickson, Eric Temple Bell, Joseph Ritt, Joseph Wedderburn, C.C. MacDuffee, Constance R. Ballantine, and Olive C. Hazlett. The latter three were all doctoral students of Dickson, the leading algebraist in the United States, a role he assumed from his mentor at the University of Chicago, E.H. Moore. On Dickson see [Fenster 2007]; on Moore, see [Parshall/Rowe 1994].

[5]This topic lingers in the background in several of the papers in [Gray/Parshall 2007].

Dickson wrote these comments about her doctoral dissertation [Sanderson 1913]: "This paper is a highly important contribution to this new field of work; its importance lies partly in the fact that it establishes a correspondence between modular and formal invariants. Her main theorem has already been frequently quoted on account of its fundamental character. Her proof is a remarkable piece of mathematics."[6]

Dickson also considered Olive Hazlett to be a particularly talented mathematician. In a letter recommending her to Edgar J. Townsend, a former student of Hilbert who became chairman of the department at the University of Illinois, Dickson wrote:

> ... She has shown more independence in research than any of our Doctors for [the] past ten years and has published perhaps a dozen excellent papers in several branches of algebra showing real originality and the ability to attack successfully quite fundamental problems. Her tested ability and her continued eagerness for research make it certain she will have a very successful career in research ....[7]

Hazlett gained an appointment at Illinois that year, and she spent the remainder of her career there. One of her students from the mid-1930s was the Hungarian-born Paul Halmos, who remembered her course on algebra. He and his fellow students thought of her as a "famous and important mathematician: she published papers and she taught advanced courses" [Halmos 1985, 45]; they learned algebra from her mainly from volume 1 of van der Waerden's *Moderne Algebra*, which within only a few years had found an international readership. Olive Hazlett had previously taught for two years at Bryn Mawr College before she was hired by Mount Holyoke in 1918. That same year, Anna Johnson Pell (Wheeler) moved from Mount Holyoke to Bryn Mawr as associate professor. In their comprehensive study [Green/La Duke 2016], the authors noted that three women were singled out as leading research mathematicians during the period 1900 to 1940: Charlotte Scott, Olive Hazlett, and Anna Pell Wheeler, all three of whom spent part of their careers at Bryn Mawr.

Emmy Noether met Hazlett in 1929, when the latter took a leave of absence to study in Göttingen. By this time, Noether was regarded as one of the leading algebraists in the world. A small, but noteworthy confirmation of the esteem she enjoyed in the United States occurred in 1931 when mathematicians at the University of Chicago requested that she send them a photo they could hang on one of the halls in Eckhardt Hall, their new headquarters built in 1930. She was eager to oblige, but turned to Hasse for help;[8] his photo of her on the ship (see Fig. 7.5) that took them to Königsberg was, in her opinion, the only decent picture of her she could send. Her problem: she had only one enlargement of it

---

[6]Cited in [Green/La Duke 2016, Entry: Sanderson]. Hazlett drew heavily on results from [Sanderson 1913] in her study [Hazlett 1921], a paper Emmy Noether reviewed for the *Jahrbuch*.

[7]Dickson to Townsend, 30 March 1925, cited in [Green/La Duke 2016, Entry: Hazlett].

[8]Noether to Hasse, 2 December 1931, [Lemmermeyer/Roquette 2006, 139].

and this was already very battered. Hasse apparently came to the rescue, and Noether's smiling face adorned the walls of Echardt Hall for many years. Chicago was long the preeminent center for algebraic research in the US, which surely made this photo a fitting acquisition. When Noether began teaching a seminar at Princeton's Institute for Advanced Study in 1934, one of those who attended was A.A. Albert, Chicago's leading algebraist.

Her strongest new American connection, though, was with the head of the mathematics program at Bryn Mawr College, Anna Pell Wheeler, who became one of Emmy Noether's good friends.[9] Born Anna Johnson, the daughter of Swedish immigrants, she grew up in Iowa and studied as an undergraduate at the University of South Dakota in Vermillion. This prairie institution was founded in 1862, thus 27 years before South Dakota became a state. Anna and her older sister Esther boarded with Alexander and Emma Pell, both of whom were immigrants from Russia. Alexander Pell was Anna's mathematics instructor, and he soon discovered her unusual talent. After graduating in 1903, she went back to Iowa for her master's degree, then went on to Radcliffe, where she took a second master's, and finally arrived in Göttingen in 1906 on a one-year fellowship from Wellesley College.[10]

Since leaving Vermillion, Anna Johnson had remained in touch with Pell, whose wife died in 1904. He traveled to Göttingen, where they married, returning to the USA in August 1907. What she knew about her husband's earlier life remains unclear, but the gist of this became known after she died [Pipes 2003]. Alexander Pell was born in 1857 in Moscow as Sergei Petrovich Degaev. He attended military schools, but then left this career behind him in 1879 to study engineering science in St. Petersburg. During this time, he became involved with the revolutionary cell known as the People's Will, a group that became famous for the assassination of Czar Alexander II in March 1881. Degaev was arrested in Odessa the following year, but escaped punishment after agreeing to collaborate with the secret police, headed by the much-feared Georgy Sudeykin. Degaev informed Sudeykin of the whereabouts of several leading members of the People's Will, including Vera Figner, and subsequent arrests nearly destroyed the military wing of the organization. Sudeykin later approached Degaev's brother, Vladimir, who informed Sergei and others. They then hatched a plan to assassinate Sudeykin and carried it out successfully in December 1883. Afterward, Degaev fled to Paris, where members of the People's Will financed his escape to America on condition that he should never set foot in Russia again. Once in the United States, he and his wife began new lives in 1891 as Alexander and Emma Pell. Six years later, Pell was awarded a Ph.D. in mathematics from Johns Hopkins University.

Pell's second wife, Anna, was 26 years younger than he, and when she returned with him to the University of South Dakota in the fall of 1907 she remained determined to finish her doctorate in Göttingen. She had begun work on a disser-

---

[9]On her career at Bryn Mawr, see [Parshall 2015, 77–80].
[10]There she attended courses offered by Hilbert, Klein, Hermann Minkowski, Gustav Herglotz and Karl Schwarzschild.

tation under Hilbert, and in the spring of 1908 she returned to Germany in order to complete it. By the end of that year, however, shortly before she planned to take her final examinations, her relationship with Hilbert had worsened to the point that she decided to return to the US without her degree.[11] In the meantime, her husband had taken a new position as assistant professor at the Armour Institute of Technology (now Illinois Institute of Technology) in Chicago. That proved convenient for Anna Pell, who in January 1909 enrolled at the University of Chicago, where she took her doctorate the following year under E.H. Moore, graduating magna cum laude. Moore accepted her thesis work in functional analysis under Hilbert, which she only had to rewrite in English. The following year, Alexander Pell suffered a stroke while teaching, so Anna taught his class for the remainder of that semester. Though he recovered, the stroke effectively ended Pell's career; his young wife cared for him until he died in 1921.

In the fall of 1911 Anna Pell became an instructor at Mount Holyoke College, where three years later she was promoted to associate professor. In 1918, she was appointed with the same rank at Bryn Mawr College, filling the vacancy left by Olive Hazlett. Five years later, Anna Pell became the first woman to deliver an invited address at a meeting of the American Mathematical Society; the second was Emmy Noether, who a decade later spoke to a large audience at Columbia University. In 1924, following Charlotte Scott's retirement, Pell became head of the mathematics department. In July 1925 she married Arthur Leslie Wheeler, a classicist who taught at nearby Princeton University. They resided together in Princeton until his death in 1932, one year before Emmy Noether's arrival in Bryn Mawr.

9.2   Emmy Noether's New Home

Well before 2 September 1933, when Emmy Noether was officially removed from the Göttingen faculty, efforts had begun to find a teaching position for her abroad. The initial idea of bringing her to Bryn Mawr came from Princeton's Solomon Lefschetz, who had met Noether in the summer of 1931 during a visit to Göttingen. Lefschetz made this suggestion to Anna Pell Wheeler some time during the spring of 1933. This was noted in a letter from July 11, written by President Marion Edwards Park to the Rockefeller Foundation, in which she inquired about supplemental funding for a foreign scholar who would be chosen by Bryn Mawr College. Park referred to a letter she had received from Edward R. Murrow, then assistant secretary of the Emergency Committee in Aid of Displaced Foreign Scholars, an organization which during the 1930s helped place some 300 Germans who had been dismissed from academic positions. Murrow's organization offered Bryn Mawr $2,000 for this purpose, assuming that the college would be able to pay

---

[11] Nothing more seems to be known about the nature of this conflict, but Anna Pell's experience was by no means unique; Hilbert could be both demanding and fickle as a mentor.

another $2,000 in salary [Kimberling 1981, 30]. Park's letter implied that Emmy Noether was the prime candidate her institution had in mind for this position.

Bryn Mawr was not the only institution, however, that had taken an interest in her case. In fact, the Principal of Somerville College in Oxford, Helen Darbyshire, had already entered into preliminary negotiations with Noether as well as with representatives of the Rockefeller Foundation in Paris. Somerville College was founded in 1879 as the the sister school of Girton College in Cambridge University. It was the first non-denominational college in Oxford. The other women's college, Lady Margaret Hall, which opened in the same year, was strictly Anglican. Initially, Emmy Noether hoped to spend at least one semester in England. President Park had in the meantime received a positive response from the Rockefeller Foundation, and on August 4, 1933, she wrote Noether to offer her an appointment at Bryn Mawr for the coming academic year as a research professor with a salary of $4,000 (the equivalent of over $76,000 today). These funds would be made available to support her research as well as consultation with advanced students. Noether's temporary position at Bryn Mawr was to be part of the Rockefeller Foundation's $1.5 million aid package for displaced scholars.

It seems likely that Emmy Noether had not anticipated receiving such a generous offer from the United States. On the other hand, her negotiations with Somerville College had progressed to the point that she felt certain to receive a firm offer from Darbyshire. In view of this attractive possibility, Noether nonchalantly suggested postponing her stay at Bryn Mawr until the academic year 1934/35. President Park did not reply immediately, and since Noether was unaware of the conditions set by the Emergency Committee and the Rockefeller Foundation, she made her plans accordingly. She described these in a letter to Richard Brauer from September 13:

> ... As for myself, I have been invited to lecture in Oxford for one term, I have chosen the time between Christmas and Easter. Subsequently, I was also offered a research professorship in Bryn-Mawr for 1933/34; I have asked to have it postponed for 1934/ 35 as I have already accepted the Oxford offer. I have no answer yet, but I think it should be all right – Bryn Mawr is a women's college, but Mitchell[12] and others are there as professors ... [Shen 2019, 57]

Richard Courant had been reluctant to leave Europe when he was placed on leave, and Emmy, too, felt disinclined to move suddenly overseas. Confusion must have reigned for a few weeks, as it was not until October 2 that President Park received definite word that Noether had accepted the invitation. The following day, the college opened its fall term, and Park thus had the opportunity to announce this important news in her convocation address.

> After a rapid fire of cables, I heard yesterday that we are to have a most distinguished foreign visitor ... in the faculty for the year, Dr. Emmy

---

[12]Howard Hawks Mitchell was an algebraist on the faculty at the University of Pennsylvania.

Noether, a member of the mathematical faculty of the University of Göttingen. Dr. Noether is the most eminent woman in mathematics in Europe and has had more students at Göttingen than anyone else in the department. With other members of the faculty, Dr. Noether was asked to resign from the University in the spring. To our great satisfaction the Institute of International Education and the Rockefeller Foundation have united in giving to the college a generous grant which makes it possible for the Department of Mathematics to invite her here for two years. Her general field is Algebra and the Theory of Numbers. Dr. Noether does not, I understand, speak English well enough to conduct a seminar at once but she will be available for consultation by the graduate students and later I trust can herself give a course. I need not say that I am delighted Bryn Mawr College is one of many American institutions to welcome the scholars whose own country has rejected them. For the time only we must believe, Germany has set aside a great tradition of reverence for the scholar and for learning. I am glad also that the college can entertain so distinguished a woman and that the students in mathematics can profit by her brilliant teaching. [Shen 2019, 58–59]

These remarks reveal that people at Bryn Mawr College, including Anna Pell Wheeler, clearly had little idea what to expect. No one even knew whether this distinguished foreign visitor could speak English. Upon her arrival, Park was pleasantly surprised to realize that Noether's "English proves to be entirely usable." Originally, Prof. Noether was to arrive on November 3 in New York, but she sent a cable on October 27 stating that her trip had been delayed due to some problems obtaining a visa. By November 7, President Park was able to inform a reporter for the Philadelphia Record that "Dr. Emmy Noether has just arrived from Germany on the Bremen after a voyage which she greatly enjoyed" [Shen 2019, 60].

The college was eager to publicize this momentous event, but at the same time took due care not to allow the press to badger their guest with questions about events in her native country. Park noted that

[she] cannot speak of German conditions during her American residence. She has a brother and many friends in Germany and she wishes herself to return for a summer. It is clear that discreet silence on her part is necessary if she is to feel at ease about her family and insure her own return. [Shen 2019, 60]

Later that month, President Park invited mathematicians from the region (including Princeton, the University of Pennsylvania, and Swarthmore) to a special lecture that Noether would give on December 15. Park also wrote to Warren Weaver in New York, inviting him to come down to see Noether "in action" on that occasion [Parshall 2015, 80].

Emmy stayed at a boarding house, run by a Mrs. Hicks, that was located only a short distance from the campus.[13] Since Anna Pell Wheeler often had students over for tea in her apartment on campus, Emmy wanted to reciprocate at her new lodgings. Many years later, Ruth Stauffer McKee remembered the scene this way:

> ...Mrs. Hicks planned a lovely tea party and Miss Noether asked Mrs. Wheeler to preside at the tea table. The setting was complete, the guest arrived, and Miss Noether beamed happily; but soon she was noticeably upset and went out to the kitchen for Mrs. Hicks. It was obvious that pouring tea, rather than being an honor, was an onerous job; and she had asked Mrs. Hicks to pour so that her good friend could enjoy herself at the party. Once again all was sunshine and light. In other words, correct an apparent problem in the simplest way. [Quinn et al. 1983, 143]

Some months later, Emmy wrote a long letter to Hasse, part of which touched on her life at Bryn Mawr:

> The people here are all very accommodating and have a natural warmth that's truly winning, even if it doesn't go very deep. You are constantly getting invitations; I've also gotten to know all kinds of interesting people who don't belong to the college. Incidentally, I am doing a seminar with three "girls" – they are only rarely called students – and a lecturer; right now they are reading van der Waerden Bd. I with enthusiasm, an enthusiasm that has them working through all the exercises – certainly not demanded by me.[14] In between I give them a little Hecke, opening chapter [Erich Hecke's *Vorlesungen über die Theorie der algebraischen Zahlen*]. For next year, however, there is – really American – an Emmy Noether Fellowship, which will probably be divided between a MacDuffee student[15] and one from Manning-Blichfeldt-Dickson;[16] the former seems to have some level. Frl. Taussky is also likely to come with a Bryn Mawr stipend; she had applied for it last year, but it wasn't granted due to lack of funds – in the past, they issued five such scholarships per year, the depression is everywhere! Finally, one of Ore's students has applied for a National Research Fellowship to come here.[17]

---

[13]In a postcard to Dirk Struik, she gave her address as The Clifton, 14 Elliott Avenue, Bryn Mawr.

[14]When volume 1 of van der Waerden's *Moderne Algebra* first came out it received an enthusiastic review from Olga Taussky and Hans Hahn (see [Koreuber 2015, 242]). One point of praise were the exercises, which they claimed could be solved by anyone who really understood the book! Taussky reviewed volume 2 on her own [Koreuber 2015, 243].

[15]Grace Shover Quinn took her doctorate at Ohio State University in 1931; her dissertation dealt with arithmetic in linear associative algebras.

[16]Marie Weiss took her Ph.D. from Stanford University in 1928.

[17]Noether to Hasse, 6 March 1934, [Lemmermeyer/Roquette 2006, 204].

That first year at Bryn Mawr, Ruth Stauffer was one of four "Noether girls" who struggled with van der Waerden's *Moderne Algebra*. Two years earlier she had earned her undergraduate degree at Swarthmore, where her adviser was Arnold Dresden. He encouraged Stauffer to do graduate work and helped her gain a graduate scholarship at Bryn Mawr for the 1931/32 academic year; she already had her master's degree when Noether arrived. Bryn Mawr then had four mathematicians on its faculty – Wheeler, Marguerite Lehr, Noether, and Gustav A. Hedlund – all of whom (except for Noether) taught undergraduate as well as graduate courses. Hedlund, an expert on topological dynamics who took his doctorate under Marston Morse at Harvard, had only recently come on board.

None of the graduate students at Bryn Mawr had ever been exposed to abstract algebra, and Ruth Stauffer remembered the shock of trying to figure out how to translate all the strange concepts in German. Miss Noether gave some simple advice: don't bother, just read the German. "That is the way our strange method of communication began. Although we students were far from conversant with the German language, it was very easy for us to simply accept the German technical terms and to think about the concepts behind the terminology. Thus from the beginning we discussed our ideas and our difficulties in a strange language composed of some German and some English" [Quinn et al. 1983, 142].

After more than four months in the United States, Noether wrote to Pavel Alexandrov with various news and some personal impressions. She had last written him in May 1933 from Göttingen and wanted to break the long silence. Apparently she had received a preliminary inquiry from Moscow about a potential professorship in algebra there, which she assumed was initiated by Alexandrov.[18] In the meantime, she had responded with a letter and postcard, but had received no reply in return. Clearly, she needed to keep all her irons in the fire, as the long-term situation remained very unclear. She now thought, though, that there might be chances of an extension after two years, to which she added:

> I hope we will meet in Princeton in the meantime! I play there once a week; it's actually a Göttingen rendez-vous! The level of the faculty there is, of course, higher than here: but insofar as the students are concerned, the difference between male and female does not seem to be as great as I originally thought. My three girls, one of whom is a lecturer, are reading v.d. Waerden enthusiastically; an enthusiasm that goes so far that they're doing all the exercises in it, an amazing feminine thing that to some extent frightens me. Next year will see an increase in the female contingent; they'll have – truly American – an Emmy Noether Fellowship for a MacDuffee student[19] who apparently knows something about algebra and number theory. In addition, Miss

---

[18] In his memorial lecture from 5 September 1935, Alexandrov mentioned various efforts on his part to offer her such a position [Alexandroff 1935, 10].

[19] Cyrus Colton MacDuffee taught at Ohio State University.

Taussky will probably get a foreign fellowship and an Ore student a local fellowship.

By the way, my English is absolutely smooth, with the result that the last miserable remains of my Russian have disappeared.[20]

Since she taught mainly by conducting dialogues, one can easily believe that by this time her English had become very fluent. Noether was obviously looking forward to the new crop of post-docs who would arrive in the coming year: Grace Shover (Quinn), Marie Weiss, and Olga Taussky (Taussky-Todd). Ruth Stauffer would stay on and eventually become Noether's only doctoral student in the United States.

Emmy had rented out her furnished apartment, and hoped she could continue doing that in the future. When she left in the fall of 1933, she imagined the trip as a long vacation, but to her own surprise, it took little time for her to feel acclimatized. As she wrote Alexandrov, "the original impossible idea of staying [in the USA] no longer seems impossible at all." She was still getting plenty of algebraic news from both inside and outside of Göttingen, and it seemed everyone was continuing to work intensely. Her sources suggested that Hasse would probably be called to Göttingen, but this was still up in the air. F.K. Schmidt was teaching in her place, and people wrote her that he was contributing a lot to maintaining her tradition. Of course, these were fragmentary impressions, "what else will happen there remains in the dark!", she added.

Emmy asked again about the present state of [Alexandroff/Hopf 1935], the now long-awaited book project. She had noticed their research announcements in the *Comptes Rendus* of the French Academy, a clear indication that their collaboration was still going strong. Speaking of algebra, she was still very excited about the prospects for work she had done with her doctoral student Chiungtze Tsen during the previous summer. Together they proved a function-theoretic analogue of Frobenius' theorem for division algebras over the real numbers [Tsen 1933]. She gave this amusing account of the background to the "Tsen" theorem:

> ... prodded by his constantly repeated questions and conjectures I got through, so that his part in this was psychologically not insubstantial, apart from the fact that he also made a few calculations. But it was surprising for everyone that there are no skew fields of finite degree over algebraic functions of a complex variable with coefficients in an algebraically closed field. Witt and others are following up intensively on that, and Artin, too, has used it for a dissertation topic.

Emmy also wrote that she had passed on Alexandrov's greetings to Anna Pell Wheeler, whom he had met during his year in Princeton with Hopf. At that time she lived in Princeton with her husband, but following his death she had moved back to Bryn Mawr. In general, Noether found the Americans she met very friendly, but also somewhat superficial. She made an exception for Wheeler,

---

[20]Noether to Alexandrov, 19 March 1934, [Tobies 2003, 105]

though, whom she found very impressive. Emmy was happy to report that Mrs. Wheeler had taken her on sightseeing trips in her car, so she had visited the Jersey coast, Manasquan, Atlantic City, and other places over the Easter holiday. She had also attended a mathematics conference held in Cambridge, and in the meantime had received invitations to visit other places in the fall, such as New Haven, Providence, and also a major conference in New York, where she was invited to deliver a lecture. She was even thinking of a trip to Canada and the Great Lakes. A woman from Toronto who had once attended lectures in Göttingen sent her an invitation, but Emmy suspected this was only a private undertaking, despite the fact that the letter came on official departmental stationery.

Noether wrote this letter shortly before she made her first trip to New York to attend the spring meeting of the American Mathematical Society, which was held on March 30–31 at Columbia University [Kline 1934]. She was not on the program, but evidently she had already received an invitation to speak at the next meeting to be held on October 27. On that occasion she spoke to a large audience about "Modern hypercomplex theories" [Kline 1935]. At the March meeting, she made some remarks after one of the talks that caused Solomon Lefschetz and A.A. Albert to sit up and take notice.[21]

Emmy Noether remained in touch with her Göttingen students as well, including Ernst Witt, Erna Bannow, and Chiungtze Tsen. She learned from them that Hasse might begin teaching there already in the coming summer semester. Since she was planning to spend a few weeks in Göttingen during early June, chances seemed very good that they could meet again. At any rate, she was keen to gain a teaching contract for Max Deuring, who had not yet habilitated, and hoped that Hasse would be able to bring that about.[22] In late April, roughly three weeks before her departure, Noether learned from F.K. Schmidt that Hasse had indeed been offered Weyl's chair in Göttingen. She then wrote to tell him how happy she was to get this news: "Now Göttingen will remain in the center! Congratulations! But actually I wish you even more luck with your latest mathematics: Hilbert's problem of class field construction and Riemann's conjecture for function fields at the same time, that's something!"[23]

Emmy's euphoria over Hasse's latest results leads one to wonder what was in his most recent letters. As Lemmermeyer and Roquette pointed out, there is no evidence Hasse thought he had found a way to solve Hilbert's twelfth problem, so what he had in mind for the first topic remains unclear. His results on the second appeared in [Hasse 1934], which represents a first step toward "Riemann's conjecture for function fields," a problem André Weil famously solved in 1941. She wrote in a similar vein to Alexandrov:

---

[21]What she had in mind was related to ideas in a no longer extant letter from Hasse, which Noether interpreted as hinting at a solution to (or perhaps progress on) Hilbert's twelfth Paris problem. Her interpretation of what Hasse claimed to have proved was, however, apparently incorrect; see [Lemmermeyer/Roquette 2006, 208].

[22]Noether to Hasse, 6 March 1934, [Lemmermeyer/Roquette 2006, 203].

[23]Noether to Hasse, 26 April 1934, [Lemmermeyer/Roquette 2006, 206].

... [Hasse] has been doing splendid things lately; he has once again solved one of the "Hilbert problems" so that not many more are left. He can, namely, in generalization of complex multiplication of elliptic functions, now construct all "class fields", i.e. all rel[ative] abelian number fields by "divisor values of abelian functions". Up until now, we had nothing here except for Hecke's results for quadratic base fields. And Hasse actually arrived at this as a by-product to the analogue of Riemann's conjecture for function fields, which he also proved. (Noether to Alexandrov, 3 May 1934, [Tobies 2003, 107])

Emmy held out high hopes for the future of algebra in Göttingen As she wrote to Alexandrov, during the past winter semester F.K. Schmidt had carried this burden all alone. Gustav Herglotz was nowhere to be seen, no doubt due to the total disruption of normal relations at the Mathematics Institute. By this time, Noether's travel plans had firmed up somewhat, and she asked Hasse again about when he would be in Göttingen. The courses at Bryn Mawr would end in mid-May, followed by exam week (which did not involve her), and then "a solemn ceremony at the end, with triple graded gowns, B.A., M.A., or Ph.D." She would visit with the Artins in Hamburg for a few days and from there travel to Göttingen, arriving in early June. "I'm especially happy for Deuring about your coming," she wrote; "I hope that the habilitation in Göttingen will go quickly; he stupidly missed that chance before going to America, and in Leipzig it seems to be dragging on because of the new regulations."

Noether regarded Deuring not only as her most talented student but also as the ideal candidate to fill her shoes. She had been very actively involved in supporting his report on algebras [Deuring 1935], a work that would replace the earlier standard study [Dickson 1927]. In fact, she had originally recommended him for this project, which took far longer than planned, primarily because of the fast-breaking developments in this field. Under normal circumstances, she would have been listed as coauthor of this work, which she went through carefully during her last visit to Göttingen in the summer of 1934.[24]

Hasse also held Deuring in high esteem and was therefore eager to appoint him. He soon came to realize, however, that the politicized student body, largely under the influence of his new colleague and co-director, Erhard Tornier, would make this very difficult. When, in April 1935, Otto Toeplitz wrote to Hasse inquiring about suitable candidates for Bonn, the latter replied: "Deuring has not yet even habilitated, though plans are now in place to do that here. I'm going to have a tough fight in this case because his type of quiet scholarly manner is not what is wanted here. Mathematically he is completely first-class" [Koreuber 2015, 249].

---

[24]For an account of Noether's role behind the scenes, see [Koreuber 2015, 245–255].

9.3   Emmy's Efforts on behalf of Fritz Noether

Over the Christmas holiday, Emmy Noether traveled to Cambridge to attend the joint annual conference of the two mathematical societies.[25] Although we know nothing about which lectures she heard or whom she spoke with on this occasion, the experience must have reminded her of the ICMs in Bologna and Zurich, if only for its sheer size with over 250 members in attendance.

During her visit, Noether stayed with two of her European friends, Dirk Jan and Ruth Ramler Struik. She first met them in 1925 when Dirk Struik was a Rockefeller fellow in Göttingen. Norbert Wiener from the Massachusetts Institute of Technology (MIT) also happened to be visiting that year, and they, too, became friends. Through Wiener's intervention, Struik received an invitation to join the MIT faculty the following year. They had thus arrived on the North American shore well before the huge wave of immigration that began in 1933. Dirk became a US citizen soon after Emmy Noether's visit, his wife only in 1939. In the meantime, they had three young daughters, so their house in Cambridge was a lively place to stay.[26]

Dirk Struik was one of the organizers for this conference, which featured various venues, alternating between MIT and Radcliffe College. He chaired a special symposium on probability theory and presumably had other duties as well, as this was a lavish affair [Richardson 1934]. Over 300 attended the evening dinner held in the Walker Memorial Building at MIT, during which the geometer Julian Lowell Coolidge presided. An entertaining after-dinner speaker, Coolidge also spoke the next day on "The rise and fall of projective geometry." Before the dinner, MIT offered an exhibition of various technical artifacts of importance for applied mathematics. For the less engineering-minded, one could instead visit the Isabella Stewart Gardner Museum, and there were afternoon tea parties at Harvard's Lowell House and Radcliffe's Agassiz House. The Boston Symphony Orchestra even gave a complimentary concert. How much of this, if any, Emmy Noether would have taken in, no one will ever likely know. Probably she spent her free time visiting with Struik's wife and her three little girls.

In fact, they had much in common, as Ruth Struik was herself an accomplished mathematician, having taken her doctorate in Prague in 1919 under Georg Pick[27] at the German Charles University, the first woman to achieve this distinction [Bečvářová 2018]. Her expertise was in foundations of geometry, whereas Dirk Struik was mainly known for his work on differential geometry and tensor analysis, much of the latter done in collaboration with J.A. Schouten in Delft. Neither thus shared strong mathematical interests with Emmy Noether, but they were never-

---

[25]At some point it was decided that these joint conferences of the American Mathematical Society (AMS) and the Mathematical Association of America (MAA) should take place in early January, and today they are often simply called the January meeting.

[26]Information on their personal lives is drawn from [Freistadt 2010], [Rowe 2018a, 379–393]; see also [Rowe 1994].

[27]Pick befriended Einstein during the academic year 1911/1912 when the latter taught in Prague. In July 1942 the Nazis sent Pick to Theresienstadt, where he died two weeks later.

theless strongly attracted to her personality and human warmth. Little is known about the circumstances of her stay in Cambridge, aside from what she briefly mentioned in a letter to Dirk Struik, written in Bryn Mawr on 25 January. There she expressed her thanks to Ruth Struik for her gracious hospitality and hoped that she had not been offended by her guest's occasional refusals to accept various kind offerings – perhaps Emmy was trying to lose weight?

Scientific couples have never had an easy time balancing their private and professional lives [Abir-Am/Outram 1987], which throughout most of the twentieth century meant that married women had no chance of pursuing a career. Saly Ruth Ramler came from a well-to-do Jewish family in Prague. As the only woman studying mathematics at the Charles University – the oldest of all German universities – Ramler encountered more than her share of incredulous stares and slighting remarks. When she presented her thesis work on affine reflections and their role in the foundations of affine geometry in Pick's seminar, the Austrian mathematician remarked afterward in perplexed disbelief: "Did you know, Miss Ramler, that you're doing axiomatics?" She turned this into her dissertation, a manuscript of some 100 pages, which her husband described as "a product of tenacity and original thinking" (he did not believe Pick gave her much help).

Dirk Struik first met Ruth Ramler in 1921 when both were attending the annual meeting of the German Mathematical Society held in Jena. Beyond their fondness for mathematics, they also shared an enthusiasm for leftist causes and Marxist ideas. They were married on Bastille Day, July 14, 1923, in the old Town Hall of Prague with its famous astronomical clock. Afterward, they lived in Delft, where Struik worked as Schouten's assistant. The newlyweds soon learned that their life was not quite as simple as it had at first seemed. As Dirk later recalled, "to our astonishment and mild amusement, we discovered after a while that we were living in sin. We were informed that the Dutch authorities did not recognize a marriage issued by the yet still young state of Czechoslovakia." Nor could they be married in Delft either, since according to Dutch law Dirk's father had to be present to sign his approval since the groom was under 30 years of age! So they made a trip to Rotterdam, the city in which Dirk Struik was born and raised, and were married a second time. In the city office, they handed over their Czech marriage certificate and tried to explain to the clerk that Ruth's name appeared there in the genitive form as Salca Ruth Ramlerova. He gave them a puzzled look, and then insisted he would copy it exactly as her name appeared in that document.

Ruth Struik was keen to move on with their lives; she disliked Schouten's Prussian manner and urged Dirk to get out from under him. This would not be easy, as academic opportunities in the Netherlands were few and far between. Their chance came, though, in April 1924 during an International Congress on Theoretical and Applied Mechanics, which took place in Delft. This event was co-organized by Jan Burgers, a friend of Dirk's since their student days in Leyden. In the meantime, Burgers had been appointed to a professorship at the Institute of Technology in Delft. Both Struiks attended the Congress, and Ruth struck on the idea of inviting some of the guests – Richard Courant, Constantin Carathéodory, Theodor von Kármán, Tullio Levi-Civita, and others – to their home. There, over an evening meal, Courant and Levi-Civita told the two of them about the Rockefeller fellowship program of the International Educational Board (IEB) and encouraged them to apply.

One month later, Struik sent a handwritten letter of application to Wickliffe Rose of the IEB in Paris with reference to this conversation in Delft. He enclosed letters of recommendation from both Levi-Civita and Courant, while briefly describing his own and his wife's academic background and accomplishments.[28] Rose's response came in the form of a perfunctory rejection in which he merely noted that the IEB could not accept applications from individuals. Twelve days later, Rose wrote to Levi-Civita, informing him that his letter recommending Struik had been read with interest by the IEB, but that he would need a letter directly from Levi-Civita in order to activate the application. He added in conclusion: "The Board would not be interested in providing a fellowship for both Dr. Struik and his wife, but if the fellowship should be provided, it would be for Professor Struik on the basis of a fellowship for a married man."[29] The Rockefeller funds were generous, but young women were almost never the beneficiaries.

They left for Rome in September and spent the next nine months there, an unforgettable time for both of them. Struik greatly admired Levi-Civita, whom he regarded as a true internationalist in the spirit of the Risorgimento. His colleague and neighbor, Federigo Enriques, was not only a prominent geometer but also one of Italy's leading historians of science. At the time, he was editor of a new Italian edition of Euclid's *Elements*, and with his customary charm he managed to convince Ruth to prepare the longest and most difficult of its 13 books, the tenth. This turned into a project that occupied her attention for many years, but with the help of Enriques' student, Maria Teresa Zapelloni, they succeeded.[30]

After this period ended, they left in June 1925 for Göttingen, since Struik's fellowship was extended to a second year, thanks to Courant's intervention. After experiencing a leisurely life in Rome's sunny and courteous surroundings, the far less easygoing atmosphere in Göttingen came as a real shock. Especially the younger lecturers loved to trade cynical remarks, packaged as sarcastic humor, a favorite target being Emmy Noether and her "boys," otherwise known as the "Unterdeterminanten," (the "minors", an algebra joke). Dirk Struik later wrote that his wife especially admired Emmy, "not only because of her way in expressing her mathematical ideas . . . but also for her courage in facing the many handicaps she had to meet as a woman – and Jewish to boot – in a masculine society in which (contrary to what we met in Rome) courtesy was not always a form of life"

---

[28]Struik to Rose, 13 May 1924, Rockefeller Archive Center, IEB Collection, 1-1, Box 60, Folder f1002, North Tarrytown, New York.

[29]Rose to Levi-Civita, 17 June 1924.

[30]The book was published in 1932 under the title *Gli Elementi d'Euclide e la Critica Antica e Moderna. Libra X.*, Bologna: Zanicelli.

[Freistadt 2010, 14–15].[31] Pavel Alexandrov delighted the Struiks with his own brand of humor, telling them how he was glad to be away from Moscow just to escape starving. It wasn't easy to be a topologist in the Soviet Union, he told them, because one had to reassure the authorities that the subject was useful for the economic recovery. So he liked to emphasize how his research field would prove useful in the textile industry (no doubt he was prepared to explain the topology of weaving patterns to anyone willing to listen). When Alexandrov learned that Dirk had purchased his winter coat with his fellowship money, he dubbed it the "paletot Rockefeller" [Rowe 2018a, 384].

Following these adventures, they returned to Delft in August 1926. At this point, Dirk was unemployed and Ruth Struik's health had begun to fail, partly due to the couple's financial insecurity. So he began searching again for other opportunities to go abroad. His brother Anton was already working as an engineer in the Soviet Union, and soon Dirk Struik received an invitation from Otto Schmidt, a mathematician and academician in Moscow, who was then planning a series of Arctic expeditions. In the meantime, Norbert Wiener had managed to arrange a visiting appointment at MIT. Both offers seemed tempting to Struik, but the first was clearly riskier, especially in view of his wife's delicate health. So after deliberating, they decided to accept the one from MIT, and in late November 1926 they boarded a ship bound for New York.

Some eight years later, Fritz Noether faced a similar situation, except that he was already 50 years old and had two grown sons. Emmy thought that there might be a chance he would be able to keep his position in Breslau, but when she wrote Dirk Struik on 25 January 1934, she knew that these efforts had now definitely come to an end. As a decorated war veteran, the Nazi's BBG stipulated that he was exempt from §3, the Aryan paragraph, which had been applied in Emmy's case. This left §4 open, which meant that Fritz had to face the far more nebulous charge of being politically unreliable. After successfully defending himself against this accusation, he lost his professorship anyway on the basis of §5, which could be applied in such cases where a person might qualify to be relieved of the stigma of being declared an enemy of the state.

As Sanford Segal pointed out, Fritz Noether's case illustrates the extent to which Hitler's government succeeded in maintaining the illusion that Germany was still a country ruled by laws [Segal 2003, 60]. The old Prussian Civil Code, enacted in 1794, was long seen as a model for modern European states, and Prussian state officials took pride in upholding its rules. After the fall of the Hohenzollern monarchy in 1918, Prussia quite astonishingly emerged as a bastion of stability during the Weimar Republic. Unlike the federal government, which went through

---

[31] In a letter to Helmut Hasse from 24 December 1930, she enclosed a mocking photomontage made by the students in Mapha, who had cut out her head from the photo Hasse had taken a few months earlier on their ship bound for Königsberg (Figure 7.5). Mapha was a student association of math and physics students, so this "present" was likely shown at a recent gathering, perhaps a Christmas party. The photomontage showed an African market with women selling their goods; one of them had the head of Emmy Noether [Lemmermeyer/Roquette 2006, 102].

13 chancellors in 14 years, the government of Prussia remained in the hands of a solid coalition of democratic parties led by the Social Democrat Otto Braun. Not until July 1932, when the reactionary Chancellor Franz von Papen drove them from power – Papen threatened to unleash the German army if Braun and his ministers did not resign peacefully – could the Nazis begin to dream of taking control of the highly efficient Prussian state bureaucracy.[32] Joesph Goebbels, who came to Berlin as Gauleiter in 1926, found it very difficult terrain for his party, though the Nazis gradually gained strength there, just as did the Communists. In the Reichstag elections on 20 May 1928 the NSDAP garnered only 1.4% of the vote in Berlin. Their road to power has been described many times as an appalling example of the failure of democratic institutions, and yet many Germans continued to believe they were still living in a constitutional state, a *Rechtsstaat*. Those who openly questioned that very assumption – Einstein, Franck, and Schouten – were few, and hence easily shouted down.

As with many other cases, Fritz Noether's problems began locally when he had to contend with radical students who demanded his dismissal. Two weeks after the promulgation of the BBG on April 7, 1933, a group of disgruntled youth protested to the Rektor in Breslau that Noether's presence contradicted the Aryan principle as well as the spirit of the new national movement. Noether voluntarily suspended his teaching for a brief time, but then took up his courses again and completed the semester. In August, students brought a new set of complaints against him, but by this time the Ministry had already linked him with various left-oriented causes that led to his dismissal on the basis of §4. Fritz Noether then appealed that decision, arguing that he had been politically inactive throughout his career. Realizing that a reversal of the decision was hopeless, he filed to be released according to the conditions stipulated in §5. One of its option called for transfer to another (presumably lower) position; another possibility allowed for the affected official to enter early retirement, which Noether requested and obtained. This also should have allowed him to qualify for regular pension benefits (had he been released on the basis of §4, his pension would have been reduced by 25%).[33]

Emmy had already spoken to Dirk Struik about her brother's situation when she wrote him on January 25. As an applied mathematician, Fritz might have under other circumstances quite easily found work somewhere in industry, but now that the United States was deeply mired in the Great Depression the outlook looked very bleak indeed. She knew that whatever pension he might draw would be far too small to feed his family, but hoped that Struik's connections might help him obtain a guest professorship in applied mathematics or mathematical physics or perhaps a scientific post in industry. Struik had apparently mentioned that John C. Slater, who chaired the physics department at MIT, might be interested in hiring him. Emmy asked him to speak with Slater and sent over Fritz's CV

---

[32] Prussian militarism clearly also suited Hitler's Nazi dictatorship, which benefited immensely from the mentality of slavish obedience (*Kadavergehorsam*).

[33] After he accepted a professorship in Tomsk, his pension from Germany was canceled [Segal 2003, 60–61]; see also [Schlote 1991].

with a list of his publications. She hoped to receive offprints of his work, which she would send later if so requested. She noted, however, that he published technical work as an employee of Siemens–Schuckert from 1920 to 1922. For more detailed information about his research, she also mentioned the possibility of writing to Richard von Mises in Istanbul, since she knew that he held Fritz's work in high esteem.

Emmy's brother had also named the Ukrainian-born applied mathematician Stephen Timoshenko, who taught at the University of Michigan, as a potentially useful contact. Perhaps Fritz knew that Timoshenko had worked for five years with Westinghouse Electric Company before he was hired in Ann Arbor and therefore assumed he still had connections with industry? Possibly they had met each other at the 1930 Stockholm Congress, but Emmy was unsure how well they knew one another. She imagined that either Struik or Slater knew Timoshenko, but otherwise she would contact him herself. Emmy was hoping for a prompt provisional reply, but when none arrived she sent Struik a postcard ten days later, on Sunday February 4, 1934. She wondered if he had in the meantime spoken with Slater or sent her brother's documents to Timoshenko. She would be going to Princeton on Wednesday and would speak with someone there about contacting Timoshenko in case that would be more appropriate.

She also asked Struik to speak with the Hungarian mathematician Otto Szász, whom Norbert Wiener had brought to MIT after he lost his position in Frankfurt. This concerned Ruth Moufang, who had taken her doctorate there in 1930 under Max Dehn. Moufang hoped to pursue an academic career in Germany, and she eventually succeeded, though it was not until 1951 that she became the first woman to be appointed to a regular professorship in mathematics. After taking her doctorate, Moufang spent a year in Rome on a post-doctoral fellowship and then taught in Königsberg during the academic year 1932/33, before returning to Frankfurt. By this time, the largely Jewish Frankfurt faculty was fast disintegrating [Bergmann/Epple/Ungar 2012, 114-132], so Dehn wrote to Emmy Noether on behalf of Moufang. Since nothing was available for her at Bryn Mawr, Emmy forwarded Dehn's letter of inquiry to Szász, now Struik's colleague at MIT. In the meantime, she wanted to know if Szász had contacted anyone at Radcliffe to see if Ruth Moufang might get a position there. None of Emmy Noether's efforts on behalf of her brother or Ruth Moufang have ever been mentioned before, no doubt because they proved entirely futile. Moufang stuck to her plan to habilitate in Frankfurt, even though Max Dehn and Ernst Hellinger were forced into retirement in 1935 and Paul Epstein chose to resign. She completed the requirements for Habilitation in 1936, but the Ministry then refused to issue the *venia legendi*, so she had to work in industry for the remainder of the Nazi era.

In another case, however, Emmy Noether's intervention proved both successful and also influential. Hans Rademacher, a left-oriented non-Jewish mathematician at the University of Breslau, was dismissed from his position already in February 1934. Originally, Hasse had recommended Rademacher for a position at the newly founded British Salem School in Scotland, led by Kurt Hahn (who had fled from the Nazis). In all likelihood, Fritz Noether (who was also in Breslau) informed his sister about Rademacher's situation, and so she spoke with mathematicians at the nearby University of Pennsylvania. On June 2, 1934, Rademacher wrote to Hasse: "Now I just found out from Miss Noether that Phil[adelphia] is completely secure for me just as soon as it becomes known there that the protest I lodged with the Ministry has been finally rejected, and that has now occurred ..." [Lemmermeyer/Roquette 2006, 214]. On 31 October 1934, Noether reported to Hasse that Rademacher was expected to arrive in Philadelphia that day. He would spend the remainder of his career at the University of Pennsylvania, where he established a major school in number theory [Siegmund-Schultze 2009, 284–286].

In that same year, Emmy Noether and Hermann Weyl established the German Mathematicians' Relief Fund, which aimed to support poorer immigrants by means of voluntary contributions from those who were better off (Weyl received a salary of $15,000 from the IAS). Initially, he and Noether asked for contributions of anywhere from 1 to 4% of the incomes from those who had been able to obtain positions in foreign countries. Among those who received some support from this fund were Ernst Hellinger, Fritz John, Hans Schwerdtfeger, and Wolfgang Sternberg [Siegmund-Schultze 2009, 197, 209].[34]

In all likelihood, Emmy received no potentially positive news from Dirk Struik regarding employment opportunities in the USA for her brother. Roughly one month later, she raised the same issue with her good friend Pavel Alexandrov. Writing him from Bryn Mawr on March 19, 1934, she mentioned that Fritz had been in touch with someone in Zurich.[35] What mainly concerned him was finding a locality that would be suitable for his two sons; the elder, Hermann, was studying chemistry in Breslau, whereas Gottfried, two years younger at age 19, was considering perhaps taking up actuarial mathematics or possibly a career in a commercial field. Emmy noted that although Gottfried was the best student in his class, he seemed not to have inherited the family's pure mathematical talent, as had once been imagined.[36] She hoped to receive news from Alexandrov right away, as she would be leaving in mid-May to meet with her brother, either in Göttingen or Breslau.[37]

---

[34]Weyl later also set up a special fund in Emmy Noether's name that helped to finance her nephew Gottfried Noether's education in the United States.

[35]At this time, the economist Fritz Demuth headed the *Notgemeinschaft deutscher Wissenschaftler im Ausland* (Emergency Association of German Scientists in Foreign Countries), which had been founded one year earlier in Zurich by the physician Philipp Schwartz. This organization soon spawned into an international network aimed at finding employment for those who had lost their positions after the Nazi government came to power. Among many other prominent émigrés, Hermann Weyl served on its advisory board. It was through the Zurich office of this organization that Fritz Noether gained a professorship in Tomsk.

[36]As it turned out, Emmy Noether's nephew later showed that he had real mathematical talent, though not in her favored direction: he became an eminent statistician in the US.

[37][Tobies 2003, 105]; it is unclear whether they met in either of these localities, but they did get together for another vacation on the Baltic.

Emmy was looking forward to this trip with great anticipation, hoping to see Hasse and others in Göttingen. Before her departure, she had received news both from her brother as well as from Alexandrov, to whom she wrote back on May 3. In the meantime, Fritz had received an offer from the Soviet Union, so it seemed "almost certain that they will go to Tomsk, a university research institute for applied mathematics and mechanics. My brother has also received favorable information for his boys" [Tobies 2003, 106].[38] Emmy was already contemplating thoughts of a trip to that remote part of the world, which would also offer her the chance for a shorter or longer stay in Moscow. Alexandrov was still trying to negotiate an algebra professorship for her there, but she made it clear that "for the time being, I don't want to commit myself to anything at all, despite the temptation." She had already made firm commitments up until the autumn of 1935, and Veblen had signaled that Princeton hoped she would stay, though it was unclear whether she would continue to commute or be given a regular position at the IAS.[39]

Memories of Moscow were now receding fast, as Emmy imagined a future life in a country she found quite fascinating. "Staying here," she wrote, "has the great advantage that – despite the dollar valuation – one can travel almost anywhere; maybe my expectations have risen in America in this regard! A 'trip abroad' doesn't mean going to the North Sea any longer! It's also a fact that English seems to have devoured all my memories of the Russian language. But at least I would still have German and English available to communicate with your students!" What she honestly hoped for the future was a wish she and Weyl both shared, namely that Princeton would eventually take the place of Göttingen for Alexandrov and other leading foreign mathematicians. If that were to "become a reality, then we could still talk about everything every day."

At the same time, Emmy's immediate thoughts focused on her brother and his family. Before her departure, she wrote to Fritz's former student Hans Baerwald and his wife Eva, reporting that her brother would be leaving with his family for Tomsk at the end of the summer, and she was planning to see them off.[40] The decision to move thousands of miles from Germany did not come easy, she noted, but Fritz Noether had received positive reports in response to his inquiries about educational opportunities for his sons. Emmy also knew, surely from Pavel Alexandrov, that the region offered excellent conditions for skiing, "no wind, strong sunshine, so that the cold weather should be easily bearable." She also had heard that a sanatorium for tuberculosis patients had been built nearby. The family "appear now to be preparing themselves learning the language, which

---

[38] Arrangements for this position in Tomsk had been negotiated by the Zurich office of the *Notgemeinschaft deutscher Wissenschaftler im Ausland.*

[39] None of these hopes for the future were to be realized, however. On the fate of Fritz Noether and his family, see [Rowe 2020b, 261–264].

[40] Emmy Noether to the Baerwalds, 15 May 1934; copy in the possession of Monica Noether.

is not easy!"[41] Since Emmy Noether had a standing offer to visit Moscow, she could already imagine visiting her brother there, even though the journey from Tomsk would take three days.

9.4 Last Visit in Göttingen

Emmy Noether arrived in Hamburg in early June and stayed for a few days with Emil and Natascha Artin. The Artins lived north of the city in Langenhorn, a region Emmy described to Hasse as like a real summer resort. Natascha was heavily pregnant with their second child, Michael, who was born on June 28. He would follow in his father's footsteps to become one of the century's premier mathematicians. Emil Artin, unlike Hasse, was a poor correspondent, so Emmy passed on his apologies to Helmut Hasse for not having written to congratulate him on his new appointment in Göttingen. She also hinted that news of the difficulties Hasse would face there had traveled to Hamburg, but without indicating any specifics. As usual, her main focus was on mathematics.

Thus, she related with excitement some news about her presentation in Artin's seminar – in which she spoke as "Noether, America" – concerning her ideas for a theory of class fields for general Galois fields. Artin had earlier tried to push something similar through without success, so both he and Emmy were now skeptical about proving a general existence theorem. Claude Chevalley happened to be in Hamburg at this time, and he offered a new hypercomplex proposal, but they quickly determined that this, too, would not work.[42] Emmy Noether was certainly very excited to be back in Germany.

Natascha recalled an exotic scene when her husband and Emmy were trying to carry on a mathematical conversation during a ride on the subway, probably before or after her talk in his seminar. As it happens, the technical vocabulary in German for concepts in class field theory could easily be mistaken for the typical political jargon of that era. In her excitement while trying to make herself heard over the surrounding noise, Emmy was drawing more and more attention to herself as the passengers heard words like *Ideale, Führer, Gruppe, Untergruppe*, and Natascha became more and more worried that they might get arrested. She was half-Jewish, which led to her husband losing his position in 1937; their flight to the United States turned out to be a precarious venture.[43]

---

[41]Herman Noether later recalled that learning Russian was a major part of the challenge they faced at Tomsk University; nevertheless, he and his brother Gottfried did very well during their three years of study there. After one year, Fritz Noether was able to teach his classes in Russian. (Notes from Herman Noether for his family, undated, courtesy of Evelyn Noether Stokvis.)

[42]Noether to Hasse, 21 June 1934, [Lemmermeyer/Roquette 2006, 209].

[43]Natascha Artin Brunswick was presumably the original source for this story, and it seems likely that she told it to Olga Taussky-Todd, one of the people interviewed by Sharon Bertsch McGrayne for her book *Nobel Prize Women in Science: Their Lives, Struggles, and Momentous Discoveries*, 2nd ed., Washington, DC: Joseph Henry Press, 1998, pp. 77, 413.

Since Emmy Noether was very reticent in writing about the political situation in Germany, it is difficult to draw a clear picture of what she knew about the specific circumstances that had radically upended mathematical life in Göttingen since her departure. Still, she almost surely knew that her former student Werner Weber, who had been Edmund Landau's assistant since 1928, was one of the ringleaders behind a protest movement that ended Landau's career. Weber may well have been radicalized by a 20-year-old student named Oswald Teichmüller, whom the faculty came to know as both a brilliant mathematician as well as a totally fanatical Nazi.[44] Teichmüller had only just turned 18 when he joined the NSDAP in July 1931 and became a member of the *Sturmabteilung* (SA), whereas Weber did not join until May 1, 1933, the exact same day as Friedrich Neumann, the new Nazi Rektor of Göttingen University and henceforth one of Weber's principal allies.

During that summer semester, Landau had followed the Dean's advice, allowing Weber to teach in his place. When the winter semester 1933/34 opened, only Landau and Gustav Herglotz remained from the original faculty. Having heard nothing to the contrary, Landau assumed he could continue teaching as usual. But when he walked over to the lecture hall a large group of some 80 students stood outside in the foyer. They parted way for him to pass, but on entering he found only one single student inside. SA members stood next to the doors, blocking entry to anyone who might have wanted to get inside. Landau withdrew to his office, soon followed by Teichmüller, who came to explain the reason for the boycott. This incident took place on November 2, at which time Landau requested that the students' case be given to him in writing [Schappacher 2000, 25]. The following day, Teichmüller delivered a lengthy statement, which included these remarks:

> ...this is not a matter of making difficulties for you as a Jew, but rather only of protecting German students in their second semester from being taught differential and integral calculus by a teacher of an entirely foreign race. I, like everyone else, do not doubt your ability to instruct suitable students of whatever origin in the purely international scientific aspects of mathematics. But I also know that many academic courses, in particular the differential and integral calculus, have at the same time a broader educational value, introducing the pupil not only to a conceptual world but also to a different intellectual sphere. Since that orientation depends very substantially on the spirit that shall be adopted, a spirit that depends, however, very essentially, following long known principles, on the racial composition of the individual, it follows that, for example, a German student should not be trained by a Jewish teacher. ...

---

[44]Teichmüller was drafted in July 1939 and took part in the invasion of Norway in 1940. He was afterward stationed in Berlin doing cryptographic work until 1942 when he was released to take up a teaching post in Berlin. After the German army suffered a crushing defeat at Stalingrad in February 1943, he volunteered for combat duty on the Eastern Front, where he was killed in action in September 1943.

So we were and remain all the more united with regard to the purpose of this action, which is mainly to restore the situation of the previous semester. Dr. Weber is prepared to substitute for you in lectures and exercises. Since the uncertainty of the previous semester no longer exists, it would not be necessary for you to speak with him before each lecture; he would rather teach the course, whether completely or in parts, on his own. We, too, would prefer that. Considering that Dr. Weber is the only one who really has to make a sacrifice, since he would double his workload in the interests of younger fellow students, while all you need do is absent yourself from the lecture course without any pecuniary or other disadvantage, I think I am offering a really easy proposal for you to accept. [Schappacher/Scholz 1992]

Two days after receiving this letter, Landau wrote to the Kurator, informing him that in view of the present situation he had no choice but to apply for early retirement.

This dramatic confrontation soon set forth shock waves throughout the international community. Richard Courant, who had recently arrived in Cambridge, informed Abraham Flexner at the Institute of Advanced Study. Shortly before this happened, he had already written him about the need to help young scientists leave Germany:

The Nazis have remained consistent only in regard to the so-called Jewish question. ... They indoctrinate a ridiculous racial theory (the basis of which is anti-Semitism) through propaganda of all kinds ... and it may easily happen that once this poisonous seed has germinated, an atmosphere much worse than that now existing will have been created. [Reid 1976, 155]

Since Emmy Noether began making regular trips to Princeton in February 1934, she presumably knew by then, at the latest, that Werner Weber had staged a successful coup by organizing a boycott of Landau's lecture course and thereby forcing him into retirement. She perhaps also knew that the Dean had named Weber acting director of the mathematical institute. In that capacity, Weber was also asked to recommend candidates for the professorship vacated by Hermann Weyl. Weber would have preferred the Nazi Udo Wegner from Darmstadt, but given the prestige attached to this chair, he named Helmut Hasse as the most suitable person to succeed Weyl as director. When Hasse was appointed in April, however, Weber reverted to his true self. Sensing he had made a terrible mistake, he began a quiet campaign to undermine Hasse's position before the latter took control of the Göttingen Mathematics Institute. The crux of this conflict stemmed from two opposing views of the university, one that favored promoting mathematical excellence in the name of German culture, the other demanding mathematical engagement for the greater cause of the national revolution. Weber came to fear that Hasse merely stood for traditional conservatism; even though he was willing to serve the Nazi state by promoting German science, Hasse failed to meet the political standards of the day because he lacked a deep commitment to Hitler's messianic mission.

In normal times, a person holding a lowly assistantship position would never have posed a threat to someone of Hasse's stature. But these, of course, were not ordinary times. Friedrich Neumann's election as the new Rektor signaled that *Gleichschaltung*, Nazi jargon for the process of institutional accommodation, now applied at Göttingen University. Indeed, Neumann, working together with the politicized *Dozentenschaft*, would ensure that all future university appointments were compatible with the principles of National Socialism. What transpired in the months that followed is a complicated story, told in considerable detail in [Segal 2003, 124–166]. Weber went to great lengths to ensure that Hasse would not have a free hand in running the institute. Landau's chair now stood vacant, and Weber's Nazi friend, Udo Wegner, was seen as a strong candidate for it. But Theodor Vahlen, who was named head of the division on higher education in the Prussian Kultusministerium, instead chose the probabilist Erhard Tornier, another ardent Nazi. Tornier was also appointed as co-director of the Göttingen Mathematics Institute, a move designed to curb Hasse's influence.

These developments were preceded by a bizarre situation that took place on May 29, when Hasse arrived in Göttingen only to discover that Weber was scheming to prevent him from becoming director of the institute. This culminated in a scene during which Weber, as acting director, refused to hand over the keys to the building. Recognizing that he was powerless to do anything in Göttingen, Hasse returned to Marburg, informed Vahlen of the present impasse, and awaited further orders from Berlin. He would only return to Göttingen on July 2, by which time Tornier had already been installed as co-director. As a backdrop to the general atmosphere, just two days earlier Hitler had unleashed a sweeping purge known as the "Night of the Long Knives," which only ended on July 2. By murdering Ernst Röhm, head of the SA, and numerous others, including several prominent conservatives, Hitler consolidated his dictatorship through lawless violence and terror. Nazism thereby revealed its true nature for all to see on the very day of Hasse's arrival, when he posted a brief notice: "Today I have taken over the leadership of the institute together with Dr. Tornier" [Segal 2003, 151].

Noether had been in Göttingen since June 7, and she soon realized that nothing was as before. Two weeks later, she wrote to Hasse in Marburg about how disappointed she was not to have found him in the institute, though by now she had apparently heard that he would soon be coming.[45] Tornier had grudgingly offered to let her use the library as a "foreign scholar," a designation that surely must have felt like adding insult to injury. She no longer had any illusions about keeping her apartment, and so had begun making plans for shipping her furniture, books, and other belongings to the United States, even though it still remained unclear whether she would be offered an extension beyond the coming academic

---

[45]Noether to Hasse, 21 June 1934, [Lemmermeyer/Roquette 2006, 209].

year. She also expressed her sadness over not being able to invite Hasse to her apartment. Realizing that if she came to Marburg this could also easily lead to problems, she suggested meeting in Bad Soden near Frankfurt, in the event Hasse was planning to visit his parents, who lived in that town. It seems unlikely, though, that they managed to meet before Hasse's arrival in Göttingen in early July.

After Hasse took up his professorship, he commuted between Göttingen and his home in Marburg. During the week, he lodged at Gebhardt's Hotel near the train station. He and Emmy took great care not to be seen together at the institute, knowing full well that Tornier and his Nazi allies would have relished the opportunity to use such evidence to attack Hasse as a friend of the old "Jewish clique" in Göttingen. Noether's major concern at this time was to ensure that Deuring could habilitate. Originally, she assumed he could do so in Leipzig, where Deuring was van der Waerden's assistant. However, as she explained in a letter to Hasse, the state of Saxony had in the meantime introduced additional requirements that would lead to unnecessary delay. Deuring had already written his postdoctoral thesis and mailed a copy to Noether. Since Hasse was spending the weekend in Marburg, Emmy wrote him on Sunday, July 15 to ask if she should bring this copy to Hasse's hotel or whether he could come to her apartment one evening.

Hasse was highly impressed with Deuring's post-doctoral thesis, as can be seen from remarks in his report.[46] Referring to its principal theorem, he wrote: "this can be seen as the first real result approaching the Riemannian conjecture and it is also of the highest significance due to the depth of the proof" [Koreuber 2015, 249]. This was written in January 1935; one month later he added a brief notice in which he underscored the importance he attached to Deuring's case:

> The initiative behind Mr. Deuring's application came from me. I have the greatest interest in bringing this excellent mathematician to Göttingen, not only because of his outstanding scientific capabilities but also because his mathematical activity continues old Göttingen traditions in an exceptional way. (*ibid.*)

It need hardly be said what traditions Hasse had in mind, but of course Emmy Noether's name could no longer be mentioned in the year 1935. Even when it came time for Deuring to deliver his qualifying lecture, Hasse advised him not to choose a topic in abstract algebra. It surely made no difference to Tornier, who in his own report made his position very clear: "I vote for Deuring's habilitation, but remark prophylactically that I would with all means oppose offering a lectureship" (*ibid.*). In light of Tornier's opposition, the Ministry refused to confer the *venia legendi* until 1938, when Deuring became a private lecturer in Jena. Thus, as Lemmermeyer and Roquette aptly noted, Emmy Noether's hopes that under Hasse "Göttingen would remain in the center" (im Mittelpunkt) could not be realized,

---

[46]Excerpts from it appear in [Koreuber 2015, 248–249]; Noether's report on his dissertation from 1930 is reproduced in [Koreuber 2015, 313–314].

especially not for her own principal field of research. Already on April 18, 1935, just four days after her death, Hasse wrote to Otto Toeplitz:

> What I find even more depressing is the fact that, on the one hand, I carry the responsibility to the mathematical world for restoring Göttingen to a place of rank, whereas, on the other hand, I am deprived of almost all influence on the personal organization due to the existing university policy regulations. This concerns not only filling vacant professorships, but also applies to lectureships, assistantships, and tutoring positions. [Lemmermeyer/Roquette 2006, 207]

Although Emmy Noether traveled to several different places during her final trip to Germany, it remains unclear whether she visited her brother and his family in Breslau. They did, however, spend some time vacationing together in a small village on the Baltic, possibly in Dierhagen, where they stayed the previous summer.[47] They were again joined by the Heisigs and the Baerwalds. Herbert Heisig and Hans Baerwald had both studied under Fritz Noether in Breslau. As a Jew, Baerwald no longer had any future in Germany, and he soon left for New York City.[48] Since Fritz Noether had acquired a car, Emmy may well have decided to travel with him and his family on their vacation to the Baltic. Their much longer trip soon afterward, from Breslau to Tomsk, nearly 3,000 miles away, was a true adventure. Since the family had been promised a large apartment, they sent over all their furniture and books, a grand piano, and even some of the older furniture from Max Noether's days in Erlangen.

Throughout the summer, Emmy Noether corresponded regularly with Olga Taussky, though often with hurried little messages, as Emmy was constantly on the move. She lectured in Marburg and Kiel, but she also wrote about plans to visit Basel and Berlin as well. Toward the end of August, Noether wrote from Magdeburg, where she was visiting friends.[49] In the meantime, Taussky learned that she would be able to postpone her fellowship at Girton College, so she could spend the coming year at Bryn Mawr with no worries about the immediate future. Noether was pleased to learn this; she had for some time been doing her best to support Taussky's future career with advice and letters of recommendation. In September, she would board a ship in Hamburg bound for a country Noether still barely knew, though she now realized that this was likely to become her new home.

Still, Emmy Noether was not one to ponder over her fate when, in fact, there was still much to do in making plans for the trip back. The week following, she would return to Göttingen so that she could organize the shipment of her furniture

---

[47]This and the following information comes from a conversation between Herman Noether and his daughter Evelyn.

[48]He and his family later moved to Cleveland, where Gottfried Noether visited them in 1939, just before he began his studies at Ohio State University.

[49]Noether to Taussky, 30 August 1933, Papers of John Todd and Olga Taussky-Todd, Box 11, Folder 11, Caltech Archives.

to the United States. In the meantime, her mind was firmly focused on what she and Frl. Taussky would experience on the other side of the ocean.

> I plan to present Chevalley's thesis in the seminar since it provides such easy access to class field theory. But I'll probably also have to use other books for the basics of Hilbert-Dedekind theory. During the previous semester, I worked through van der Waerden I and Mitchell's number theory[50] with the four seminar students. This was something between a seminar and tutorial and it worked quite well, so with that you will be able to help! In Princeton, where I teach once a week, I'll do the same thing only from a slightly more advanced point of view.[51]

9.5 Lecturer at Princeton's Institute for Advanced Study

Emmy Noether's association with Princeton's Institute for Advanced Study (IAS) first began in February 1934 when she started making weekly visits, usually on Tuesdays, to deliver lectures there. She suggested a bit of the flavor of what this was like in a letter to Hasse:

> ... I've been lecturing once a week in Princeton – at the Institute and not at the "men's"-university, which does not admit anything female, whereas Bryn Mawr had more male lecturers than females, so is only exclusive regarding students. At the beginning, I started with representation modules and groups with operators; this winter Princeton will for the first time get treated algebraically, and quite thoroughly. Weyl also lectures on representation theory, although he will soon switch to continuous groups. Albert ... lectured before Christmas on something hypercomplex in the style of Dickson, together with his "Riemannian matrices". Vandiver, who is also on "leave of absence", is lecturing on number theory for the first time in an eternity at Princeton.[52] And after I had given my survey on class field theory in the Mathematics Club, von Neumann ordered twelve copies of Chevalley as a textbook (Bryn Mawr shall also get some!). I was also told that your Lecture Notes will be translated into English, hopefully now in sufficiently many copies – I had pestered the people about this already in the fall. My audience consists essentially of research fellows, along with Albert and Vandiver, but I noticed that I have to be careful; these people are used to explicit

---

[50]Howard H. Mitchell, who was Oswald Veblen's first doctoral student at Princeton, taught a number theory course at Bryn Mawr before Noether's arrival. Since he never wrote a textbook on number theory, this reference suggests that he probably gave Noether some notes for a continuation of his earlier course.

[51]Noether to Taussky, 30 August 1933, Papers of John Todd and Olga Taussky-Todd, Box 11, Folder 11, Caltech Archives.

[52]Harry Vandiver was a professor at the University of Texas; in 1931 he won the Cole Prize of the AMS for his work on the Fermat conjecture.

computations, and I have already driven some of them away! The University and Flexner Institute together have more than sixty professors and those who want to be; even if Princeton tries to draw many of them, all these research fellows are a sign of academic unemployment.[53]

Noether's assessment of the general employment picture in the US was certainly accurate, even though Princeton served as a springboard for many, including her collaborator Richard Brauer. As concerns the physical surroundings for mathematicians, Emmy's claim not to be at the "men's"-university stands in need of clarification. During its early years of operation, the IAS was housed in Fine Hall along with the mathematics department. Or, to be more precise, they were together in the old Fine Hall, today called Jones Hall, which was built in 1930 and originally named in memory of the department's founding chairman, Dean Henry B. Fine.[54] Henry Fine had been one of Felix Klein's first American doctoral students, taking his degree in Leipzig in 1885. He was by no means a leading researcher, but he played a pivotal role in making Princeton one of the three leading universities for mathematics in the United States (the other two were Chicago and Harvard) [Parshall/Rowe 1994, 438 451]. Shortly after Dean Fine's death, in 1929, Jones and his niece provided for the erection of a mathematics building in his memory, and, what was rarer, an endowment for its upkeep. Feeling that "nothing is too good for Harry Fine," Jones said that the building to bear his name should be a place which "any mathematician would be loath to leave." The finished building featured a spacious wood-paneled library, common rooms, and faculty studies. It also contained a locker room with shower bath for faculty wishing to use the then-nearby tennis courts; this amenity inspired these lines about the department chairman in the Faculty Song:

> "He's built a country-club for Math
> Where you can even take a bath."

Loath as the mathematicians were to leave, in 1969 the increased size of the department compelled them to move into the new Fine Hall, leaving their marks on the old one – mathematical formulas and figures in the leaded design of the windows, and Einstein's famous remark over the fireplace in what is now the lounge: *Raffiniert ist der Herr Gott, aber Boshaft ist Er nicht* (God is subtle, but He is not malicious). Oswald Veblen found that saying so delectable, he decided to have it engraved there for posterity.[55]

Emmy Noether had the pleasure of reconnecting with Richard Brauer on her weekly trips to Princeton. During the previous year, Brauer taught at the University of Kentucky in Lexington, after which Hermann Weyl invited him to spend

---

[53]Noether to Hasse, 6 March 1934, [Lemmermeyer/Roquette 2006, 204].

[54]When the new Fine Hall was completed in 1969, the old one was renamed in honor of its donors, Thomas D. Jones and his niece Gwethalyn Jones.

[55]This saying alludes to Einstein's skepticism regarding the new orthodoxy in quantum physics, as reflected in his remark that "God doesn't throw dice."

the academic year 1934/35 at Princeton's Institute for Advanced Study. His principal duty that year was to prepare the text for Weyl's lectures on Structure and Representation of Continuous Groups [Weyl 1934/35]. They also wrote the paper [Brauer/Weyl 1935], in which they constructed an n-dimensional representation for spinors. During this academic year, Brauer and his wife Ilse (who was also a mathematician) became very close friends with Emmy Noether.

Hasse and Noether remained in steady contact after she returned to the United States. Emmy reported to him that some 200 people from New York and surrounding colleges came to hear her speak at the recent AMS meeting held at Columbia. Probably she visited the Courants during this trip, as she wrote further that Richard "feels very well in New York – he lives 40 minutes away by car in a rural area not far from the beach – and is becoming more human."[56] About her own lectures at the IAS, Noether gave Hasse this glimpse of her teaching activity:

> There are a number of interested people in Princeton this year; I am doing a seminar on class field theory, although it's mainly a lecture course with occasional [active] participation by other people. For the time being, however, we are still stuck in Galois theory; but next time Ms. Taussky, who occasionally comes along [from Bryn Mawr], will present some simple number-theoretic examples. She held a rehearsal here, and I'm doing the same, but tailored for women, i.e. the girls replace what they lack in self-reliance with an uncanny diligence – this year there are two others besides Miss Taussky here on a scholarship.
> [Lemmermeyer/Roquette 2006, 212–213]

Coming from Vienna, interrupted by her year in Göttingen, Olga Taussky already knew a number of the mathematicians she would meet again in Princeton. If, for Emmy Noether, the gatherings in Fine Hall were like a "Göttingen rendezvous," they seemed to Olga more like a joint meeting of mathematicians from Göttingen and Vienna. Wilhelm Magnus, her co-worker on the Hilbert edition was there, as of course were Veblen and Weyl. She had met Richard Brauer at the 1930 conference in Königsberg, and she had done coursework in Vienna with Walter Mayer, who came to the IAS as Einstein's assistant. Taussky even knew John von Neumann from talks he had given in Vienna and Göttingen. Little wonder that her trips with Emmy were the highlight of her stay in the United States. Their "rehearsals" for the girls took place on a Monday, the day before they took the train to Princeton Junction, a 3-hour trip. Olga Taussky could not afford to go every week, partly because her European fellowship did not include a waiver of tuition fees. Still, she went whenever she could, even just for the chance to "talk mathematics" with Noether, who in such a situation could be a very good listener. During the two-year interim since she left Göttingen, Taussky became interested in topological algebra after reading a paper from 1932 by Lev Pontryagin. When she told Noether about this on the way to Princeton, Emmy

---

[56]Noether to Hasse, 31 October 1934, [Lemmermeyer/Roquette 2006, 213]. Courant apparently did not attend the meeting at Columbia.

Figure 9.2: Bryn Mawr College, June 1935: (l. to r.) Marie Weiss, Ruth Stauffer, and Grace Shover, May 1935 (Auguste Dick Papers, 13-1, Austrian Academy of Sciences, Vienna)

afterward introduced her to the topologist James Alexander, who then told her that Nathan Jacobson was very familiar with Pontryagin's paper. This soon led to a joint publication by Jacobson and Taussky on locally compact rings.

Ruth Stauffer (Fig. 9.2) was now in her second year of studies under Emmy Noether, who teased her in a friendly way about her name. It reminded her of Werner Stauffacher, a character in Schiller's play William Tell, so she kept assuring her student that she was surely of Swiss origin. Many years later, Ruth Stauffer McKee told an anecdote about a typical local excursion with her fellow students, led by their irrepressible teacher. In those days, one did not need to wander far from the Bryn Mawr campus to reach the countryside, and so they began walking across an open field, and Stauffer noticed that they were headed straight for a rail fence:

> Miss Noether was immersed in a mathematical discussion and went merrily along, all of us walking at a good clip. We got closer and closer to the fence. I was apparently the weak sister, concerned mostly in how we would handle the fence. For those of us in our twenties it would be no problem but, from my point of view, however would this "old lady," fiftyish, handle the fence? On we went right up to the fence and without missing a word in her argument she climbed between the rails and on we went [Quinn et al. 1983, 144].

Grace Shover Quinn also remembered trips to Philadelphia and Princeton in Mrs. Wheeler's car, and another trip to Swarthmore College, a little more than a half-hour's drive away [Quinn et al. 1983, 140]. Like Bryn Mawr and Haverford College, all nearby, Swarthmore was a traditional Quaker school going back to the nineteenth century.[57] On this trip, the car must have been quite full with the driver, Emmy, Ilse and Richard Brauer, and Grace Shover. They went to Swarthmore to visit Isaac and Charlotte Schoenberg, another émigré couple. Isaac Jacob Schoenberg was a Rumanian Jew who took his doctorate under Issai Schur in Berlin. He then worked under Edmund Landau in Göttingen on analytic number theory. Through Landau he got a position at Hebrew University in Jerusalem, before coming to the US on a Rockefeller fellowship in 1930, the year he married Landau's daughter, Charlotte. From 1933 to 1935 he had a fellowship at the IAS. His sister Irma Wolpe was a pianist married to Hans Rademacher; they met in India when he was on leave at the Tata Institute in Bombay.

Olga Taussky-Todd later recalled that Emmy Noether often felt irritated during their year together at Bryn Mawr, and since those feelings were very often expressed in German, she became a kind of sounding board for Emmy. At the same time, she sometimes became annoyed with Noether for directing quite a lot of petty criticism at her, beginning with her Austrian accent and green felt hat with a feather. Emmy said it reminded her of someone they both knew – no doubt Gustav Herglotz – when he wore Lederhosen. So Taussky gave the hat to Ruth Stauffer, since she knew how much Ruth loved it [Taussky 1981, 87]. Taussky was now 28 and had taken her doctorate four years earlier under Furtwängler, so she came to resent Noether's motherly ways, all the while not knowing about Emmy Noether's many worries, which "der Noether" kept very much to herself. Taussky called her a "tough guy," and felt that Noether expected the same of her; Emmy surely recognized that Olga Taussky had the talent and ambition to go far in her career.

For Emmy Noether, oddly enough, the key to success for women in mathematics was marriage. She colluded with Ilse Brauer in trying to line up husbands for her four students, but with no success. As Taussky saw her in retrospect, Noether was

> very naive and knew very little about life. She saw women as being protected by their families and even admitted to me that she gave young men preference in her recommendations for jobs so they could start a family. She asked me to understand this, but, of course, I did not. [Taussky 1981, 91]

There was also a clash in their mathematical styles. Noether was trying to teach these young women cutting-edge research based on Hasse's mimeographed lecture notes on class field theory (Noether had reviewed this work in *Zentralblatt* in 1933). Clearly one needed a solid background in algebraic number theory to understand such advanced material, so she tasked Taussky with presenting lectures

---

[57]Haverford admitted men only until the 1970s, whereas Swarthmore was from its founding co-educational; Bryn Mawr remains today a women's college.

designed to fill in some of those gaps. Not surprisingly, Taussky chose to base her presentations on Hilbert's *Zahlbericht*, which did not please Noether at all. For her, this was just computing, not the kind of conceptual algebraic approach she wanted to see, and Emmy Noether was not one to hide her opinions on mathematical matters. This put real pressure on Olga Taussky, who later made light of the situation in a little poem. Her inspiration for it came from a witty piece by Wilhelm Busch about a bird stuck in a tree, a poor creature that realized it was about to be devoured by a cat. Emmy Noether would have surely enjoyed it, but Taussky was too bashful to show her this delightful creation (first in German [Dick 1970/1981, 1970: 34] and then in English [Dick 1970/1981, 1981: 84–85]):

> Es steht die Olga vor der Klasse,
> sie zittert sehr und denkt an Hasse.
> Die Emmy kommt von fern herzu,
> mit lauter Stimm', die Augen gluh.
> Die Trepp hinauf und immer höher
> kommt sie dem armen Mädchen näher.

> Die Olga denkt: weil das so ist
> und weil mich doch die Emmy frisst,
> so werd' ich keine Zeit verlieren,
> werd' keine Algebra studieren
> und lustig rechnen wie zuvor.
> Die Olga, dünkt mir, hat Humor.[58]

> Olga stands outside the class room
> with wrinkled brow and in deep gloom.
> Emmy, from far away comes along
> with a firm step and feeling strong.
> She climbs upstairs with a great swirl
> and gets quite close to the poor girl.

> Now Olga thinks: Of hope there is no ray
> and Emmy scolds me anyway.
> Merrily I will compute some more
> and algebra I will ignore.
> It makes me think that Olga had humor.

Emmy thought a lot about Hasse, too, and in one of her letters to him she described her latest ideas for pursuing class field theory by combining transcendental methods with Chevalley's new algebraic ideas. Hasse discussed this with Ernst Witt and both agreed that if one needed analytical methods to found class field theory – which seemed at that time to be the case – then one should point that powerful cannon straight at the fortress of the theory. More precisely, "one

---

[58]Nach Wilhelm Busch, "Es sitzt ein Vogel auf dem Leim."

should use Witt's remounting of Käte Hey's cannon and then fire a return shot on the classical class field theory as was done on your 50th birthday."[59]

Noether was very enthused about her Princeton lectures during this second year, as can be seen from a letter to Hasse from 28 November. She reported that Morgan Ward was diligently studying Hasse's lecture notes from her seminar.[60] Zariski was also attending and had begun to immerse himself in the arithmetic theory of algebraic functions. Before Christmas time, Noether sent a witty message to Hasse, who answered in turn:

> Thank you for your content-rich letter! I can well imagine that with the stormy weather and the hopeless classes you spend your time dancing on tiptoe, that you are exhausted, cannot enjoy the primitive meals, but are still happy to be there anyway. I am very honored that you need me and are willing to sacrifice money, love, beer, sleeping, eating, and vacationing, whereas I feel ashamed since you really have nothing for which to thank me.[61]

Emmy sorely missed Hasse and the life she once had in Göttingen. All things considered, though, she was adjusting remarkably well to her new situation; still, she was worried about her brother and his family, not to mention the state of her own deteriorating health. She had undergone an operation the previous summer, during which her physician discovered a large uterine fibroid. Her original plan was to have it removed when she returned to Göttingen in the summer of 1935. Presumably, Emmy Noether had informed Anna Pell Wheeler of this, but in any event, on the advice of another physician, Noether decided to undergo this second operation in Bryn Mawr.

On top of these worries, her employment situation was still unsettled, and she still had no news about this as the year 1934 came to an end. Since her initial two-year appointment would expire the following summer, Anna Pell Wheeler contacted Oswald Veblen to inform him that Bryn Mawr had no funding available to keep her on the faculty. Veblen then wrote to Abraham Flexner on 13 December 13 1934, alerting him to the impasse:

> The professors of the Institute would be quite willing to recommend a small grant-in-aid for a year or two, especially in view of the fact that Miss Noether has been lecturing here during the last two years. And this might help to bridge the gap in case it is necessary to make

---

[59]Noether to Hasse, 31 October 1934, and Hasse to Noether, 19 November 1934, [Lemmermeyer/Roquette 2006, 212–215], referring to his paper for her 50th birthday [Hasse 1933]; Hey's cannon is a reference to Käte Hey's doctoral dissertation from 1927, written under Emil Artin, see [Dumbaugh/Schwermer 2018].

[60]Ward took his doctorate at Caltech under Eric Temple Bell in 1928 with a dissertation entitled "The Foundations of General Arithmetic"; he joined the faculty there the following year. Olga Taussky met him again at a number theory conference held at Caltech, and he was later instrumental in arranging her appointment as the first woman on its faculty.

[61]Hasse to Noether, 17 December 1934, [Lemmermeyer/Roquette 2006, 219].

temporary arrangements for a couple of years longer. In view of Miss Noether's unique position in the world – the only woman mathematician of the first order – it ought to be possible to find some persons or group of people who would make it possible for Bryn Mawr to keep her permanently. [Shen 2019, 61]

Flexner had doubts about such a short-term commitment and also felt that the IAS had already done a great deal already for German scholars. He was concerned not to create the impression that his institution was overlooking Americans in order to help unfortunate foreigners.

These years were, indeed, particularly difficult ones for young American mathematicians, one of whom was Nathan Jacobson, a Polish Jew whose family immigrated to the USA in 1918. Jacobson took his doctorate at Princeton under Wedderburn in 1934, spent the year 1934/35 at the IAS, where he attended Noether's lectures on class field theory, and then, following her death, was hired by Anna Pell Wheeler to teach at Bryn Mawr for one year. After that he was awarded a one-year fellowship from the National Research Council to do post-doctoral work with Abraham Adrian Albert and Leonard Dickson at the University of Chicago. By the spring of 1937, he was again looking for a regular position. As he later recalled: "... this was the depths of the Great Depression. Salaries declined in some instances and there were very few new positions. Moreover, for the new Jewish Ph.D.'s the situation was further aggravated by anti-Semitism that was prevalent, especially in the top universities – the only ones that had any interest in fostering research" [Niven 1988, 220].[62]

Sidestepping these issues, Veblen was eventually able to secure a $1,500 grant, while he continued soliciting larger donations for a "permanent commitment on the part of the Institute." As he put it, Noether was not merely unique as a "woman mathematician," she offered the Institute an opportunity to capitalize on the brain-drain from Göttingen by supporting "one of the most important scientists" displaced by the events in Germany. A parallel effort was undertaken by the Dutch-American mathematician Arnold Dresden at nearby Swarthmore College.[63] After meeting in Philadelphia with Jacob Billikopf, executive director of the Federation of Jewish Philanthropies, Dresden contacted several leading mathematicians who wrote letters of support [Kimberling 1981, 34–36].

On 31 December, Lefschetz wrote:

As the leader of the modern algebra school, she developed in recent Germany the only school worthy of note in the sense, not only of isolated work, but of very distinguished group scientific work. In fact, it is no

---

[62] Jacobson was hired by the University of North Carolina at Chapel Hill; he then taught at Johns Hopkins from 1943 to 1947 before joining the faculty at Yale University.

[63] Dresden came to Swarthmore from the University of Wisconsin in 1927. He was recruited by President Frank Aydelotte in order to initiate an honors program in mathematics that would later serve as a model for other colleges and universities throughout the U.S. Aydelotte was a member of the board of the IAS and later served as its director, which may have played a role in Dresden's initiative.

exaggeration to say that without exception all the better young German mathematicians are her pupils. Were it not for her race, she would have held a first rate professorship in Germany and we would have no occasion to concern ourselves with her. She is the outstanding refugee German mathematician brought to these shores and if nothing is done for her, it will be a true scandal.

Norbert Wiener from MIT, writing on 2 January, was just as emphatic:

Miss Noether is a great personality; the greatest woman mathematician who has ever lived; and the greatest woman scientist of any sort now living, and a scholar at least on the plane of Madame Curie. Leaving all questions of sex aside, she is one of the ten or twelve leading mathematicians of the present generation in the entire world and has founded what is certain to be the most important close-knit group of mathematicians in Germany – the Modern School of Algebraists. Even after she was deprived of her position in Germany on account of her sex, race and liberal attitude, numbers of students (men as well as women) continued to meet at her rooms for mathematical instruction. Of all the cases of German refugees, whether in this country or elsewhere, that of Miss Noether is without doubt the first to be considered.

G.D. Birkhoff from Harvard also encouraged this undertaking, which he implied should not be as costly as in other cases of similarly prominent mathematicians.

Miss Noether ... is generally regarded as one of the leaders in modern Algebraic Theory. Within the last ten or fifteen years she and her students in Germany have led the way much of the time. It is not too much to say that, since Sonia Kovalevski, she is the only woman mathematician of high absolute rank. Thus it is an opportunity for us to have her in this country and I hope very much that you will succeed in your efforts. Her continued presence at Bryn Mawr is sure to be a stimulus to everyone interested in modern Algebra in this country.

As far as the desirable arrangements to be made in her case are concerned I find myself in general agreement with my colleague Professor Wiener. I might mention here the fact that at least when I was last in Germany her salary there was not large, and also that as far as undergraduate work is concerned, she will be probably of no use at Bryn Mawr.

Somewhat complicated negotiations then began between Bryn Mawr and the two main agencies that had financed Noether's position until this time, the Emergency Committee and the Rockefeller Foundation. Warren Weaver, from the Rockefeller Foundation, was eventually persuaded to support Noether for the coming academic year. He saw no prospect for a long-term appointment at Bryn Mawr, due primarily to her exclusive interest in research-level mathematics. That being the case, her supporters focused on the possibility of creating a permanent position for her at the IAS. On 28 February, Veblen wrote to Flexner about

> ...the possibility that this might become a permanent commitment on the part of the Institute. There is no doubt that, apart from the uniqueness of her position as a woman mathematician, she is quite obviously one of the most important scientists who have been displaced by the events in Germany. Therefore even a permanent commitment could be nothing but creditable to the Institute. [Shen 2019, 62]

Weaver decided that the Rockefeller Foundation would make an exception for Noether with regard to its standard policy on academic refugees. He was thus preparing to authorize her appointment at Bryn Mawr for another two years. As matters turned out, however, she would never finish her original two-year term.

9.6 Emmy Noether's Tragic Death

Grace Shover Quinn later recalled some of the events from the last two weeks of Emmy Noether's life. In late March, the college was on spring break, so the dorms were closed and everyone went their separate ways. Olga Taussky went to Atlantic City, and Emmy took Grace Shover along to visit her there on Sunday, March 31. Exactly one week later, Shover visited Noether again, this time with a friend from Germany, who enjoyed chatting with the famous mathematician in the apartment she rented south of the Lancaster Pike. She had all her furniture shipped from Germany, which made her feel quite at home, especially with her massive desk. The next day, Monday April 8, Mrs. Wheeler summoned the four girls to let them know that Miss Noether would be entering the hospital that day for removal of a uterine tumor. The operation was performed that Wednesday. Probably Noether had informed Anna Pell Wheeler that a similar operation had been performed the summer before in Göttingen, and that a second such procedure had been scheduled there for the coming summer. In any event, she was advised not to postpone this second operation any longer.

Her four students visited their teacher the day before the operation and planned to see her again on Saturday, but on that day they were told she was not feeling well enough to see visitors. In fact, she had been doing fine, but her condition suddenly worsened that very day. The next afternoon, when they were in their dormitory rooms, Ruth Stauffer was called to the telephone and learned from Anna Pell Wheeler that Emmy Noether had passed away. Various stories about the cause of death apparently went around shortly thereafter, but her physician, Dr. Brooke M. Anspach, wrote to President Park the day after Noether's death informing her that the patient was suffering from a cerebral lesion which led to a ruptured blood vessel. After she became unconscious on early Sunday morning, her temperature rose to 108 degrees. The operation no doubt aggravated this condition, but in the opinion of Dr. Anspach, if the tumor had not been removed it alone would have caused her death [Shen 2019, 62].

Emmy Noether knew that her condition was very serious; she may have even realized that her life was at risk when she entered the hospital on April 8, 1935. Before leaving home, she made a list of items she owned that should be distributed to those dear to her in the event she did not survive the operation. Grace Shover received a necklace and Olga Taussky a brooch and one of Dedekind's books from his own personal library.[64] In her last letter to Helmut Hasse, Noether described Ruth Stauffer's work.[65] She wrote nothing about going in for an operation and only mentioned that she was unsure whether she would visit Göttingen again in the summer, but if so then not until the end of June. She had to be present at the commencement ceremony to present her student to the president, who would then place the doctoral hood on Ruth Stauffer's head. It was not to be, and so Richard Brauer had to step in for Emmy.[66]

Fritz Noether received the sad news in Berlin on Monday, April 15. He immediately sent a telegram to Hasse as a member of the executive committee of the German Mathematical Society. On the same day, Hasse also received a telegram from Hermann Weyl. He answered by requesting that Weyl purchase a wreath in the name of the Göttingen mathematicans for the funeral ceremony on Wednesday. This was a traditional Quaker service that took place in President Park's living room. During the Bryn Mawr Symposium, Grace Shover Quinn recalled the scene on that somber occasion [Quinn et al. 1983, 141]:

> We heard music played softly by a string ensemble in a nearby room. A closed black box along the side of the room reminded us of the loss of our beloved professor. Four eulogies were scheduled: Mrs. Wheeler represented Miss Noether's American colleagues; Richard Brauer, speaking in German, her German colleagues; Ruth Stauffer, her American students; and Olga Taussky, her foreign students.

During the morning chapel service the day afterward, Emmy Noether's colleague Marguerite Lehr recalled the excitement and confusion that preceded her arrival in November 1933:

> ... there was much discussion and rearrangement of schedule, so that graduate students might be free to read and consult with Miss

---

[64]During the next year, Shover taught at the Shipley School in Bryn Mawr, a position she had taken the previous winter with the intention of continuing her studies with Noether. In the summer of 1936, she attended the International Mathematical Congress in Oslo, where she met Fritz Noether. Before the Congress, she traveled to Göttingen and attended lectures by Helmut Hasse.

[65]Noether to Hasse, 7 April 1935, [Lemmermeyer/Roquette 2006, 221].

[66]After receiving her degree in June 1935, Ruth Stauffer taught mathematics for one year at the Bryn Mawr School in Baltimore while studying with Oscar Zariski at the Johns Hopkins University. In 1937 she married George McKee, who like her grew up in Harrisburg. She worked for some 30 years as a statistician for a research agency attached to the Pennsylvania State Legislature.

Noether until she was ready to offer definitely scheduled courses. For many reasons it seemed that a slow beginning might have to be made; the graduate students were not trained in Miss Noether's special field – the language might prove a barrier – after the academic upheaval in Göttingen the matter of settling into a new and puzzling environment might have to be taken into account. When she came, all of these barriers were suddenly non-existent, swept away by the amazing vitality of the woman whose fame as the inspiration of countless young workers had reached America long before she did. ...

Professor Brauer in speaking of Miss Noether's powerful influence professionally and personally among the young scholars who surrounded her in Göttingen said that they were called the Noether family, and that when she had to leave Göttingen, she dreamed of building again somewhere what was destroyed there. We realize now with pride and thankfulness that we saw the beginning of a new "Noether family" here. To Miss Noether her work was as inevitable and natural as breathing, a background for living taken for granted; but that work was only the core of her relation to students. She lived with them and for them in a perfectly unselfconscious way. She looked on the world with direct friendliness and unfeigned interest, and she wanted them to do the same. She loved to walk, and many a Saturday with five or six students she tramped the roads with a fine disregard for bad weather. Mathematical meetings at the University of Pennsylvania, at Princeton, at New York began to watch for the little group, slowly growing, which always brought something of the freshness and buoyancy of its leader. [Quinn et al. 1983, 144–145]

Fritz Noether and his cousin in Mannheim, Otto Noether, both expressed their gratitude to Bryn Mawr for providing Emmy with a second home. Fritz also wrote to Professor Wheeler on 23 May 1935:

I know also from other reports that she felt at home at Bryn Mawr, and Bryn Mawr had become an absolute substitute for what she had to give up in her homeland. I also see from your report, how well you all know her, her idiosyncrasies, and her main traits – the unbreakable optimism which she evidently held till the last hours. Painful as the thought is to us all that she is no longer here with us, the greatest satisfaction remains that she herself kept living and working in her ideas until the moment that her thinking stopped, without her becoming aware of it. [Shen 2019, 64]

Those who knew Emmy Noether best were her fellow Germans in exile, in particular her former colleague in Göttingen, Hermann Weyl. On April 26, Weyl delivered [Weyl 1935], his well-known address for a memorial service held at Goodhart Hall on which occasion an urn containing Noether's ashes was interred in the

(a) Locality of Emmy Noether's Ashes   (b) Gravestone Marker

Figure 9.3: Final Resting Place at Bryn Mawr College (Courtesy of Qinna Shen)

Cloisters at Bryn Mawr (Fig. 9.3). In our introduction, we mentioned some of his remarks, but he also spoke about their time together before they left for the United States.

> When I was called permanently to Göttingen in 1930, I earnestly tried to obtain from the Ministerium a better position for her, because I was ashamed to occupy such a preferred position beside her whom I knew to be my superior as a mathematician in many respects. I did not succeed, nor did an attempt to push through her election as a member of the Göttingen Gesellschaft der Wissenschaften. Tradition, prejudice, external considerations, weighted the balance against her scientific merits and scientific greatness, by that time denied by no one. In my Göttingen years, 1930–1933, she was without doubt the strongest center of mathematical activity there, considering both the fertility of her scientific research program and her influence upon a large circle of pupils. [Weyl 1968, 3: 432]

At the very end of this address, Weyl's words took on an emotional edge, but otherwise he spoke about Emmy Noether respectfully, almost impersonally. Perhaps he felt his main task was to give an American audience at least a glimpse of the remote world she came from, the small city of Erlangen during the quiet era before the Great War. A few of his less flattering remarks have also been repeatedly cited, which might easily lead to the impression that he had limited sympathy for her. This was hardly the case, as should be evident from the preceding chapter. Until quite recently, in fact, it was completely overlooked that Hermann Weyl had also spoken at the more intimate funeral service held on April 17. On that private occasion, speaking in German, Weyl expressed far more deeply how he felt just a few days after her death. In but a few words he captured the profound tragedy of her passing, fully recognizing that the world they had shared only a short time ago was now gone forever. Only her mathematics would remain.[67]

The hour has struck, Emmy Noether, for us to say goodbye to you forever. Your death will move many deeply; none more so than your beloved brother Fritz, who living almost half a world away cannot be here and can only speak his last farewell through my mouth. These flowers that I place for you on the coffin are from him. We bow in acknowledgment of his pain, which we are not entitled to put into words.

But I feel it is an obligation at this hour to express the feelings of your German colleagues, those who are here and those in your homeland who have been loyal to our goals and to you personally. And at your grave I would like to do so in our mother tongue – the language you felt in your heart and in which you conceived your thoughts – which remains sacred to us no matter what power reigns on German soil. You will rest in foreign soil, in the earth of this great hospitable country that offered you a place of work after your own country closed itself off from you. At this moment we feel the urge to thank America for what it has done for German science in the past two pressing years, and especially to thank Bryn Mawr College, which was both happy and proud to include you among its teachers, and rightly so. Because you were a great female mathematician, I have no hesitation in calling you the greatest that history can record. Through your work algebra has acquired a new face. With many Gothic letters, you have inscribed your name indelibly on its pages. Perhaps no one has contributed as much as you in transforming axiomatic thinking, which before was only used to elucidate the logical foundations, into a powerful tool for concrete, forward-moving research. Amongst your predecessors in algebra and number theory probably Dedekind came closest to you.

When I think of your being in this hour, two traits, above all, appear as most important. The first is the primal, productive power of your mathematical thinking. Like an over-ripe fruit, it seemed to burst through the shell of your humanity. You were the instrument and vessel of the intellect that broke forth from within you. There were no tender considerations, it was always the matter at hand that de-

---

[67]Translated from [Roquette 2007, 19–20].

manded attention. There was nothing gentle and well balanced in your being; you were not clay, formed into a harmonious shape by the hands of God, but rather a chunk of primordial human rock into which he breathed creative spirit. The power of your genius seemed to transcend the bounds of your sex, which is why we in Göttingen, in awed mockery, often spoke of you in the masculine form as "der Noether." And yet you were a motherly woman with the warm heart of a child. You gave your pupils full and abundant intellectual support, and they gathered around you like chicks under the wings of a large mother hen. You loved them, cared for them, and lived in close communion with them.

And that is the second feature of your nature that I think is most significant: your heart knew no malice; you did not believe in evil, indeed, it never occurred to you that evil played a role in human life. This never impressed me more than in the last stormy summer of 1933 that we spent together in Göttingen. In the midst of the terrible struggle, collapse and rupture that raged around us in all factions, in a sea of hatred and violence, of fear and despair and burdensome worry – you went your usual way, pondering mathematical problems with the same zeal as before. When you were denied access to a lecture hall in the institute, you gathered your students in your own apartment; you remained friends with those who wore the brown shirt, you never doubted their honesty for a moment. Unconcerned about your own fate, fearless and open and conciliatory as always, you went your own way. Many of us believed that an enmity had been unleashed that could not be pardoned; none of this touched your soul. You were happy to return to Göttingen last summer, lived and worked in the circle of like-minded German mathematicians, as if everything had remained as before; you planned to do the same this summer. You truly deserve the wreath that the Göttingen mathematicians have asked me to lay on your grave.

We do not know what death is. But isn't it a comforting thought to imagine that after this earthly life our souls might recognize each other again, and how your father's soul would then greet you? Has any father ever found a daughter who was such a great and independent successor? – In the midst of your full creative power you were suddenly torn from us; your sudden departure still stands like a flash of lightning written on our faces. But your memory will long remain alive in science and among your students, friends and colleagues; you have ensured this through your work and your personality. Farewell, Emmy Noether, you great mathematician and great woman. Your perishable remains shall pass away, that which is imperishable we want to preserve.

Bibliography

[Abir-Am/Outram 1987] Abir-Am, Pnina G., Outram, Dorinda, eds.: *Uneasy Careers and Intimate Lives: Women in Science, 1789–1979*, New Brunswick: Rutgers University Press.

[Albert 1930] Albert, A.A.: New results in the theory of normal division algebras, *Transactions of the American Mathematical Society* 32: 171–195.

[Alexandroff 1927] Alexandroff, Paul: Über stetige Abbildungen kompakter Räume, *Mathematische Annalen* 96: 555–571.

[Alexandroff 1928] Alexandroff, Paul: Über den allgemeinen Dimensionsbegriff und seine Beziehungen zur elementaren geometrischen Anschauung, *Mathematische Annalen* 98: 617–635.

[Alexandroff 1932a] Alexandroff, Paul: *Einfachste Grundbegriffe der Topologie*, Berlin: Springer.

[Alexandroff 1935] Alexandroff, Paul: In Memory of Emmy Noether, (Memorial Address, 5 September 1935), in [Noether 1983, 1–11].

[Alexandroff 1976] Alexandroff, Paul: Einige Erinnerungen an Heinz Hopf, *Jahresbericht der Deutschen Mathematiker-Vereinigung* 78: 113–125.

[Alexandrov 1979/1980] Alexandrov, Paul: Pages from an autobiography, *Russian Mathematical Surveys*, 34(6): 267–302; 35(3): 315–358.

[Alexandroff/Hopf 1935] Alexandroff, Paul und Hopf, Heinz: *Topologie* Berlin: Julius Springer.

[Barrow-Green 1997] Barrow-Green, June: *Poincaré and the Three Body Problem*. History of Mathematics. 11. Providence, RI: American Mathematical Society.

[Beery et al. 2017] Beery, Janet L. et al., eds.: *Women in Mathematics: Celebrating the Centennial of the Mathematical Association of America*, New York: Springer

[Bečvářová 2018] Bečvářová, Martina: Saly Ruth Struik, 1894–1993, *Mathematical Intelligencer* 40(4): 79–85.

© The Author(s), under exclusive license to Springer Nature Switzerland AG 2020
D. E. Rowe, M. Koreuber, *Proving It Her Way*, https://doi.org/10.1007/978-3-030-62811-6

[Bergmann/Epple/Ungar 2012] Bergmann, Birgit, Epple, Moritz, Ungar, Ruti, eds.: *Transcending Tradition: Jewish Mathematicians in German-Speaking Academic Culture*, Heidelberg: Springer.

[Blumenthal 1935] Blumenthal, Otto: Lebensgeschichte, in [Hilbert 1935, 388–429]

[Bongiorno/Curbera 2018] Bongiorno, Benedetto und Curbera, Guillermo: *Giovanni Battista Guccia: Pioneer of International Cooperation in Mathematics*, New York: Springer.

[Bourbaki 1960] Bourbaki, Nicolas: *Éléments d'historie des mathématique*, Paris: Hermann.

[Brauer/Hasse/Noether 1932] Brauer, Richard, Helmut Hasse und Emmy Noether: Beweis eines Hauptsatzes in der Theorie der Algebren, *Journal für die reine und angewandte Mathematik* 167: 399–404; reprinted in [Noether 1983, 630–635].

[Brauer/Weyl 1935] Brauer, Richard, and Weyl, Hermann: Spinors in n dimensions, *American Journal of Mathematics* 57: 425–449.

[Brewer/Smith 1981] Brewer, James W. and Smith, Martha K.: *Emmy Noether. A Tribute to her Life and Work*, New York: Marcel Dekker.

[Brieskorn/Purkert 2018] Brieskorn, Egbert und Purkert, Walter: *Felix Hausdorff: Gesammelte Werke, Biographie*, Bd. IB, Heidelberg: Springer.

[Brieskorn/Scholz 2002] Brieskorn, Egbert und Scholz, Erhard: Zur Aufnahme mengentheoretisch-topologischer Methoden in die Analysis Situs und geometrische Topologie, in [Hausdorff 2002, 70–75].

[Brill 1923] Brill, Alexander: Max Noether, *Jahresbericht der Deutschen Mathematiker-Vereinigung*, 32: 211–233.

[Brill/Noether 1894] Brill, Alexander und Noether, Max: Die Entwicklung der Theorie der algebraischen Functionen in älterer und neuer Zeit, *Jahresbericht der Deutschen Mathematiker-Vereinigung*, 3: 107-566.

[Bruns-Wüstefeld 1997] Bruns-Wüstefeld, Alex: *Lohnende Geschäfte: Die „Enjudung" der Wirstschaft am Beispiel Göttingens*, Hannover: Fackelträger Verlag.

[Castelnuovo/Enriques/Severi 1925] Castelnuovo, Guido, Enriques, Federigo, Severi, Francesco: Max Noether, *Mathematischen Annalen*, 93: 161–181.

[Clebsch/Lindemann 1876] Clebsch, Alfred und Lindemann, Ferdinand: *Vorlesungen über Geometrie von Alfred Clebsch. Bearbeitet und herausgegeben von Ferdinand Lindemann*, Leipzig: Teubner.

[Cohn/Taussky 1978] Cohn, Harvey and Taussky, Olga: *A classical invitation to algebraic numbers and class fields. With two appendices by Olga Taussky: "Artin's 1932 Göttingen lectures on class field theory" and "Connections between algebraic number theory and integral matrices"*, New York: Springer.

[Cooke 1984] Cooke, Roger: *The Mathematics of Sonya Kovalevskaya*, New York: Springer.

[Corry 2004] Leo Corry: *Modern Algebra and the Rise of Mathematical Structures*, 2nd ed., Basel: Birkhäuser.

[Corry 2017] Corry, Leo: Steht es alles wirklich schon bei Dedekind? Ideals and factorization between Dedekind and Noether, *In Memoriam Richard Dedekind (1831–1916)*, Katrin Scheel, Thomas Sonar, Peter Ullrich, eds., Münster: Verlag für wissenschaftliche Texte und Medien, pp. 134–159.

[Corry/Schappacher 2010] Corry, Leo, and Schappacher, Norbert: Zionist Internationalism through Number Theory: Edmund Landau at the Opening of the Hebrew University in 1925, *Science in Context* 23(4): 427–471.

[Courant/Hilbert 1924] Courant, Richard and Hilbert, David: *Methoden der mathematischen Physik*, Bd. 1, Berlin: Springer.

[Curtis 2007] Curtis, Charles W.: Emmy Noether's 1932 ICM Lecture on Non-commutative Methods in Algebraic Number Theory, in [Gray/Parshall 2007, 199–219].

[Dahms 1999] Dahms, Hans Joachim: Die Universität Göttingen 1918 bis 1989: Vom ‚Goldenen Zeitalter' der Zwanziger Jahre bis zur ‚Verwaltung des Mangels' in der Gegenwart, *Göttingen: Von der preussischen Mittelstadt zur südniedersächsischen Grossstadt 1866-1989*, Rudolf von Thadden und Jürgen Trittel, Hrsg., Göttingen: Vandenhoeck & Ruprecht, 395–456.

[Dahms 2008] Dahms, Hans Joachim: Einleitung *Die Universität Göttingen unter dem Nationalsozialismus*, 2te Ausgabe, H. Becker, H-J. Dahms, C. Wegeler, Hrsg., München: K.G. Saur, S. 29–73.

[Dedekind 1877] Dedekind, Richard: Festschrift zur Saecularfeier des Geburtstages von Carl Friedrich Gauss dargebracht vom Herzoglichen Collegium Carolinum zu Braunschweig: Über die Anzahl der Ideal-Klassen in den verschiedenen Ordnungen eines endlichen Körpers, Braunschweig 1877. Vieweg; reprinted in [Dedekind 1930–32, 1: 105–157].

[Dedekind 1894a] Dedekind, Richard: Supplement XI. Über die Theorie der ganzen algebraischen Zahlen, von *Vorlesungen über Zahlentheorie von P.G. Lejeune Dirichlet*, in [Dedekind 1930–32, 3: 1–313].

[Dedekind 1894b] Dedekind, Richard: Zur Theorie der Ideale, *Nachrichten der königlichen Gesellschaft der Wissenschaften zu Göttingen*, Mathematisch-physikalische Klasse, 272–277, in [Dedekind 1930–32, 2: 43–49].

[Dedekind 1895] Dedekind, Richard: Über die Begründung der Idealtheorie, *Nachrichten der königlichen Gesellschaft der Wissenschaften zu Göttingen*, Mathematisch-physikalische Klasse, 106–113, in [Dedekind 1930–32, 2: 50–58].

[Dedekind 1930–32] Dedekind, Richard: *Gesammelte mathematische Werke*, 3 Bde., Emmy Noether, Robert Fricke, Øystein Ore), Braunschweig: Vieweg.

[Dedekind/Weber 1882] Dedekind, Richard und Weber, Heinrich: Theorie der algebraischen Functionen einer Veränderlichen, *Journal für die reine und angewandte Mathematik* 92: 181–290.

[Deuring 1935] Deuring, Max: *Algebren*, Berlin: Springer.

[Dick 1970/1981] Dick, Auguste: *Emmy Noether*, Engl. trans. H.I. Blocher, Boston: Birkhäuser.

[Dickson 1910] Dickson, Leonard E.: Hensel's Theory of Algebraic Numbers, *Bulletin of the American Mathematical Society* 17(1): 23–36.

[Dickson 1923a] Dickson, Leonard E.: A New Simple Theory of Hypercomplex Integers, *Journal des Mathématiques Pures et Appliquées* 2: 281–326.

[Dickson 1923b] Dickson, Leonard E.: *Algebras and their Arithmetics*, Chicago. University of Chicago Press.

[Dickson 1927] Dickson, Leonard E.: *Algebren und ihre Zahlentheorie*, Zürich: Orell Füssli.

[Dubreil 1982] Dubreil, Paul: L'algèbre, en France, de 1900 à 1935, *Cahiers du séminaire d'histoire des mathématiques* 3: 69–81.

[Dubreil 1983] Dubreil, Paul: Souvenirs d'un boursier Rockefeller 1929–1931, *Cahiers du séminaire d'histoire des mathématiques* 4: 61–73.

[Dumbaugh/Schwermer 2018] Dumbaugh, Della and Schwermer, Joachim: Käte Hey and Margaret Matchett – Two Women PhD Students of Emil Artin, [Beery et al. 2017, 51–66].

[Eckes/Schappacher 2016] Eckes, Christophe and Schappacher, Norbert: Dating the Gasthof Vollbrecht Photograph, published online at https://opc.mfo.de

[Edwards 1990] Edwards, Harold M.: Takagi, Teiji, *Dictionary of Scientific Biography*, vol. 18, Supplement II, New York: Charles Scribner's Sons, 890–892.

[Einstein 1998] *The Collected Papers of Albert Einstein. The Berlin years: Correspondence, 1914–1918*, vol. 8, Robert Schulmann, A.J. Kox, Michel Janssen, József Illy, eds., Princeton: Princeton University Press.

[Fenster 2007] Fenster, Della Dumbaugh: Research in Algebra at the University of Chicago: Leonard Eugene Dickson and A. Adrian Albert, in [Gray/Parshall 2007, 179–197].

[Fenster/Parshall 1994] Fenster, Della Dumbaugh, and Parshall, Karen Hunger: Women in the American Mathematical Research Community: 1891–1906, *The History of Modern Mathematics, vol. 3: Images, Ideas, and Communities*, David E. Rowe and Eberhard Knobloch, Boston: Academic Press, pp. 229–261.

[Ferreirós 2007] Ferreirós, José: *Labyrinth of Thought. A History of Set Theory and Its Role in Modern Mathematics*, 2nd ed., New York: Springer.

[Frei 1985] Frei, Günther, Hrsg.: *Der Briefwechsel David Hilbert–Felix Klein (1886–1918)*, Arbeiten aus der Niedersächsischen Staats- und Universitätsbibliothek Göttingen, Bd. 19, Göttingen: Vandenhoeck & Ruprecht.

[Frei/Roquette 2008] Frei, Günther u. Roquette, Peter, Hrsg.: *Emil Artin und Helmut Hasse – Die Korrespondenz 1923-1934*, Göttingen: Universitätsverlag.

[Freistadt 2010] Freistadt, Margo: *Dirk Jan and Saly Ruth Struik*, San Francisco, private printing.

[Furtwängler 1929] Furtwängler, Philipp: Beweis des Hauptidealsatzes für die Klassenkörper algebraischer Zahlkörper, *Abhandlungen aus dem Mathematischen Seminar der Universität Hamburg* 7: 14–36.

[Goldstein/Schappacher/Schwermer 2007] Goldstein, Catherine, Schappacher, Norbert, Schwermer, Joachim, eds.: *The Shaping of Arithmetic after C.F. Gauss's Disquisitiones Arithmeticae*, Heidelberg: Springer.

[Goodstein 2020] Goodstein, Judith: Olga Taussky-Todd, *Notices of the American Mathematical Society* 67(3): 678-687.

[Gordan 1868] Gordan, Paul: Beweis, dass jede Covariante und Invariante einer binären Form eine ganze Function mit numerischen Coefficienten einer endlichen Anzahl solcher Formen ist, *Journal für die reine und angewandte Mathematik* 69: 323–354.

[Gordan 1885/1887] Gordan, Paul: *Vorlesungen über Invariantentheorie*, 2 Bde., Georg Kerschensteiner, Hrsg., Leipzig: Teubner.

[Gray 2006] Gray, Jeremy: A History of Prizes in Mathematics, *The Millennium Prize Problems*, Carlson, J.; Jaffe, A. and Wiles, A. eds., Providence, RI: American Mathematical Society.

[Gray 2018] Gray, Jeremy: *A History of Abstract Algebra. From Algebraic Equations to Modern Algebra*, Cham: Springer Nature Switzerland.

[Gray/Parshall 2007] Gray, Jeremy and Parshall, Karen, eds., *Episodes in the History of Modern Algebra (1800–1950)*, History of mathematics series, vol. 32, Providence, RI: American Mathematical Society.

[Green/La Duke 2009] Green, Judy, and Jeanne La Duke: *Pioneering Women in American Mathematics: The Pre-1940 PhD's*, Providence, RI: American Mathematical Society.

[Green/La Duke 2016] Green, Judy, and Jeanne La Duke: Supplementary Material for Pioneering Women in American Mathematics: The Pre-1940 PhD's. Last updated January 15, 2016. http://www.ams.org/publications/authors/books/postpub/hmath-34-PioneeringWomen.pdf. Accessed 11 March 2020.

[Grell 1927] Grell, Heinrich: Beziehungen zwischen den Idealen verschiedener Ringe, *Mathematische Annalen* 97: 490–523.

[Grüttner/Kinas 2007] Grüttner, Michael, und Kinas, Sven: Die Vertreibung von Wissenschaftlern aus den deutschen Universitäten 1933–1945, *Vierteljahrshefte für Zeitgeschichte* 55: 123–186.

[Halmos 1985] Halmos, Paul: *I Want to Be a Mathematician*, New York: Springer.

[Hashagen 2003] Hashagen, Ulf: *Walther von Dyck (1856–1934). Mathematik, Technik und Wissenschaftsorganisation an der TH München* (Boethius, Texte und Abhandlungen zur Geschichte der Mathematik und der Naturwissenschaften, Bd. 47), Stuttgart: Steiner.

[Hasse 1926] Hasse, Helmut: Bericht über neuere Untersuchungen und Probleme aus der Theorie der algebraischen Zahlkörper. I: Klassenkörpertheorie, *Jahresbericht der Deutschen Mathematiker-Vereinigung* 35: 1–55.

[Hasse 1927] Hasse, Helmut: Existenz gewisser algebraischer Zahlkörper, *Sitzungsberichte der Preußischen Akademie der Wissenschaften*, 229–234.

[Hasse 1928] Hasse, Helmut: Besprechung von L.E. Dickson, Algebren und ihre Zahlentheorie, *Jahresbericht der Deutschen Mathematiker-Vereinigung* 37: Abt. 2, 90–97.

[Hasse 1930] Hasse, Helmut: Die moderne algebraische Methode, *Jahresbericht der Deutschen Mathematiker-Vereinigung* 39: 22–34.

[Hasse 1931] Hasse, Helmut: Über $\wp$-adische Schiefkörper und ihre Bedeutung für die Arithmetik hyperkomplexer Zahlsysteme, *Mathematische Annalen* 104: 495–534.

[Hasse 1932a] Hasse, Helmut: Zu Hilberts algebraisch-zahlentheoretischen Arbeiten, in [Hilbert 1932, 528–535].

[Hasse 1932b] Hasse, Helmut: Theory of cyclic algebras over an algebraic number field, *Transactions of the American Mathematical Society* 34: 171–214.

[Hasse 1933] Hasse, Helmut: Die Struktur der R. Brauerschen Algebrenklassengruppe über einem algebraischen Zahlkörper, insbesondere Begründung des Normenrestsymbols und die Herleitung des Reziprozitätsgesetzes mit nichtkommutativen Hilfsmitteln, *Mathematische Annalen* 107: 731–760.

[Hasse 1934] Hasse, Helmut: Abstrakte Begründung der komplexen Multiplikation und Riemannsche Vermutung in Funktionenkörpern, *Abhandlungen aus dem Mathematischen Seminar der Hamburgischen Universität* 10: 325–348.

[Hausdorff 2002] Hausdorff, Felix: *Felix Hausdorff: Gesammelte Werke, Grundzüge der Mengenlehre*, Bd. II, Egbert Brieskorn et al., Hrsg., Heidelberg: Springer.

[Hausdorff 2008] Hausdorff, Felix: *Felix Hausdorff: Gesammelte Werke, Deskriptive Mengenlehre und Topologie*, Bd. III, Ulrich Felgner et al., Hrsg., Heidelberg: Springer.

[Hausdorff 2012] Hausdorff, Felix: *Felix Hausdorff: Gesammelte Werke, Korrespondenz*, Bd. IX, Walter Purkert, Hrsg., Heidelberg: Springer.

[Hawkins 2000] Hawkins, Thomas: *Emergence of the Theory of Lie Groups: An Essay in the History of Mathematics, 1869–1926*, New York: Springer.

[Hawkins 2013] Hawkins, Thomas: *The Mathematics of Frobenius in Context*, New York: Springer.

[Hazlett 1921] Hazlett, Olive C.: New proofs of certain finiteness theorems in the theory of modular covariants, *Transactions of the American Mathematical Society* 22: 144–57.

[Hazlett 1924] Hazlett, Olive C.: Two Recent Books on Algebras, *Bulletin of the American Mathematical Society* 30(5–6): 263–270.

[Hensel 1908] Hensel, Kurt: *Theorie der algebraischen Zahlen*, Leipzig: Teubner.

[Hensel 1913] Hensel, Kurt: *Zahlentheorie*, Berlin und Leipzig: Göschen.

[Hentschel 1996] Hentschel, Klaus, ed.: *Physics and National Socialism: An Anthology of Primary Sources*, Basel: Birkhäuser.

[Herbrand 1932a] Herbrand, Jacques: Théorie arithmétique des corps de nombres de degré infini. Extensions algébriques finies de corps infinis, *Mathematische Annalen* 106: 473–501.

[Herbrand 1932b] Herbrand, Jacques: Zur Theorie der algebraischen Funktionen. (Aus Briefen an E. Noether), *Mathematische Annalen* 106: 502.

[Hermann 1926] Hermann, Grete: Die Frage der endlich vielen Schritte in der Theorie der Polynomideale. Unter Benutzung nachgelassener Sätze von K. Hentzelt, *Mathematische Annalen* 95: 736–788.

[Hesseling 2003] Hesseling, Dennis E.: *Gnomes in the Fog. The Reception of Brouwer's Intuitionism in the 1920s*, Basel: Birkhäuser.

[Hilbert 1890] Hilbert, David: Über die Theorie der algebraischen Formen, *Mathematische Annalen* 36: 473–534.

[Hilbert 1893] Hilbert, David: Über die vollen Invariantensysteme, *Mathematische Annalen* 42: 313–373.

[Hilbert 1897] Hilbert, David: Die Theorie der algebraischen Zahlkörper, *Jahresbericht der Deutschen Mathematiker-Vereinigung*, 4: 175–546; reprinted in [Hilbert 1932, 63–363].

[Hilbert 1900] Hilbert, David: Mathematische Probleme. Vortrag, gehalten auf dem Internationalen Mathematikerkongreß zu Paris, 1900. *Nachrichten der Königlichen Gesellschaft der Wissenschaften zu Göttingen. Math.-phys. Klasse*, 253–297; reprinted in [Hilbert 1935, 290–329].

[Hilbert 1902] Hilbert, David: Über die Theorie der relativ-Abel'schen Zahlkörper, *Acta Mathematica* 26: 99–131.

[Hilbert 1915] Hilbert, David: Die Grundlagen der Physik (Erste Mitteilung), *Nachrichten der königlichen Gesellschaft der Wissenschaften zu Göttingen*, Mathematisch-physikalische Klasse, 395–407.

[Hilbert 1932] Hilbert, David: *Gesammelte Abhandlungen*, Bd. 1, Berlin: Springer.

[Hilbert 1933] Hilbert, David: *Gesammelte Abhandlungen*, Bd. 2, Berlin: Springer.

[Hilbert 1935] Hilbert, David: *Gesammelte Abhandlungen*, Bd. 3, Berlin: Springer.

[Hilbert/Cohn-Vossen 1932] Hilbert, David und Cohn-Vossen, Stephan: *Anschauliche Geometrie*, Berlin: Springer.

[Hofmann/Betsch 1998] Hofmann, K. H., und G. Betsch: Hellmuth Kneser: Persönlichkeit, Werk und Wirkung, Preprint Nr. 2009 TU Darmstadt.

[Honda 1975] Honda, Kin-ya: Teiji Takagi: A Biography, *Commentarii mathematica Universitatis Sancti Pauli* 24: 141–167.

[Hurewicz/Wallman 1948] Hurewicz, Witold and Wallman, Henry: *Dimension Theory*, Princeton: Princeton University Press.

[Jentsch 1986] Jentsch, Werner: Auszüge aus einer unverøöffentlichten Korrespondenz von Emmy Noether und Hermann Weyl mit Heinrich Brandt, *Historia Mathematica* 13: 5–12.

[Kaplan 1991] Kaplan, Marion A.: *The Making of the Jewish Middle Class: Women, Family, and Identity in Imperial Germany*, Oxford: Oxford University Press.

[Kaufholz-Soldat 2019] Kaufholz-Soldat, Eva: A Divergence of Lives: Zur Rezeptionsgeschichte von Sofja Kowalewskaja (1850–1891) um die Wende vom 19. zum 20. Jahrhundert, Dissertation, Johannes Gutenberg Universität Mainz.

[Kimberling 1981] Kimberling, Clark: Emmy Noether and her Influence, in [Brewer/Smith 1981, 3–61].

[Klein 1872] Klein, Felix: *Vergleichende Betrachtungen über neuere geometrische Forschungen* (Das „Erlanger Programm"), Erlangen: A. Deichert.

[Klein 1904] Klein, Felix: Über die Aufgaben und die Zukunft der philosophischen Fakultät, *Jahresbericht der Deutschen Mathematiker-Vereinigung* 13: 267–276.

[Klein 1921–23] Klein, Felix: *Gesammelte Mathematische Abhandlungen*, 3 vols., Berlin: Julius Springer.

[Klein 1922] Klein, Felix (für die Redaktion der *Mathematischen Annalen*): Widmung: Max Noether, *Mathematischen Annalen*, 85: 7–9.

[Klein/Sommerfeld 1910] Klein, Felix und Sommerfeld, Arnold: *Über die Theorie des Kreisels*, Heft 4: Die technischen Anwendungen der Kreiseltheorie, Leipzig: Teubner.

[Kleiner 2007] Kleiner, Israel: *A History of Abstract Algebra*, Boston: Birkhäuser.

[Kline 1934] Kline, J.R.: The March meeting in New York, *Bulletin of the American Mathematical Society* 40(5): 353–357.

[Kline 1935] Kline, J.R.: The October meeting in New York, *Bulletin of the American Mathematical Society* 41(1): 1–4.

[Kneser 1926] Kneser, Hellmuth: Die Topologie der Mannigfaltigkeiten, *Jahresbericht der Deutschen Mathematiker-Vereinigung* 34: 1–14.

[Koenigsberger 1904] Koenigsberger, Leo: *Carl Gustav Jacob Jacobi*, Leipzig: Teubner.

[Koenigsberger 1919] Koenigsberger, Leo: *Mein Leben*, Heidelberg. Digitale Ausgabe, Gabriele Dörflinger, 2015, http://archiv.ub.uni-heidelberg.de/volltextserver/19762/1/leben.pdf

[Koreuber 2015] Koreuber, Mechthild: *Emmy Noether, die Noether-Schule und die moderne Algebra. Zur Geschichte einer kulturellen Bewegung*, Heidelberg: Springer.

[Koreuber 2021] Koreuber, Mechthild, Hrsg.: *Wie kommt das Neue in die Welt? Interdisziplinäres Symposium aus Anlass des 100-jährigen Jubiläums der Habilitation Emmy Noethers*, Heidelberg: Springer.

[Kosmann-Schwarzbach 2006/2011] Kosmann-Schwarzbach, Yvette: *Les Théorèmes de Noether. Invariance et lois de conservation au XXe siècle, avec une traduction de l'article original Invariante Variationsprobleme*, Éditions de l'École Polytechnique, deuxième édition, révisée et augmentée; *The Noether theorems: Invariance and conservation laws in the twentieth century*, Sources and Studies in the History of Mathematics and Physical Sciences, Springer.

[Krull 1935] Krull, Wolfgang: *Idealtheorie*, Berlin: Springer.

[Lasker 1905] Lasker, Emanuel: Zur Theorie der Moduln und Ideale, *Mathematischen Annalen* 60: 19–116.

[Lemmermeyer 2007] Lemmermeyer, Franz: The Development of the Principal Genus Theorem, in [Goldstein/Schappacher/Schwermer 2007, 529–561].

[Lemmermeyer/Roquette 2006] Lemmermeyer, Franz u. Roquette, Peter: *Helmut Hasse und Emmy Noether - Die Korrespondenz 1925–1935*, Göttingen: Universitätsverlag.

[Lemmermeyer 2018] Lemmermeyer, Franz: David Hilbert: Die Theorie der algebraischen Zahlkörper, *Jahresbericht der Deutschen Mathematiker-Vereinigung* 120(1): 41–79.

[Lorenat 2020] Lorenat, Jemma: "Actual Accomplishments in This World": The Other Students of Charlotte Angas Scott, *Mathematical Intelligencer* 42(1): 56–65.

[Macaulay 1916] Macaulay, F.S.: *Algebraic Theory of Modular Systems*, Cambridge Tracts in Mathematics and Mathematical Physics, no. 19, Cambridge : Cambridge University Press.

[McLarty 2005] McLarty, Colin: Poor Taste as a Bright Character Trait: Emmy Noether and the Independent Social Democratic Party, *Science in Context* 18(3): 429–450.

[McLarty 2012] McLarty, Colin: Theology and its discontents, the origin myth of modern mathematics, *Circles Disturbed : The Interplay of Mathematics and Narrative*, Apostolos Doxiadis and Barry Mazur, eds., Princeton: Princeton University Press, pp. 105–129.

[McLarty 2017] McLarty, Colin: The Two Mathematical Careers of Emmy Noether, [Beery et al. 2017, 231–252].

[McMurran/Tattersall 2017] McMurran, Shawnee L. and Tattersall, James J.: Fostering Academic and Mathematical Excellence at Girton College, 1870–1940, in [Beery et al. 2017, 3–35].

[Mehrtens 1987] Mehrtens, Herbert: Ludwig Bieberbach and "Deutsche Mathematik," in *Studies in the History of Mathematics*, Esther Phillips, ed., Washington: The Mathematical Association of America, pp. 195–241.

[Merzbach 1983] Merzbach, Uta C.: Emmy Noether: Historical Contexts, in [Srinivasan/Sally 1983], pp. 161–171.

[Neumann 2007] Neumann, Olaf: Divisibility Theories in the Early History of Commutative Algebra and the Foundations of Algebraic Geometry, in [Gray/Parshall 2007, 73–105].

[Niven 1988] Niven, Ivan: The Threadbare Thirties, *A Century of Mathematics in America*, vol. 1, Peter Duren, et al., eds., Providence, RI: American Mathematical Society, pp. 209–229.

[Noether 1908] Noether, Emmy: Über die Bildung des Formensystems der ternären biquadratischen Form, *Journal für die reine und angewandte Mathematik* 134: 23–90; reprinted in [Noether 1983, 31–99].

[Noether 1913] Noether, Emmy: Rationale Funktionenkörper, *Jahresbericht der Deutschen Mathematiker-Vereinigung* 22: 316–319; reprinted in [Noether 1983, 141–144].

[Noether 1915] Noether, Emmy: Körper und Systeme rationaler Funktionen, *Mathematische Annalen* 76: 161–191; reprinted in [Noether 1983, 145–180].

[Noether 1916] Noether, Emmy: Die allgemeinsten Bereiche aus ganzen transzendenten Zahlen, *Mathematische Annalen* 77: 103–128; reprinted in [Noether 1983, 195–220].

[Noether 1918a] Noether, Emmy: Invarianten beliebiger Differentialausdrücke, *Nachrichten der Königlichen Gesellschaft der Wissenschaften zu Göttingen*, Mathematisch-Physikalische Klasse, 37–44; reprinted in [Noether 1983, 240–247].

[Noether 1918b] Noether, Emmy: Invariante Variationsprobleme, *Nachrichten der Königlichen Gesellschaft der Wissenschaften zu Göttingen, Mathematisch-Physikalische Klasse*, 1918: 235–257; reprinted in [Noether 1983, 248–270].

[Noether 1919] Noether, Emmy: Die arithmetische Theorie der algebraischen Funktionen einer Veränderlichen in ihrer Beziehung zu den übrigen Theorien und zu der Zahlkörpertheorie, *Jahresbericht der Deutschen Mathematiker-Vereinigung* 28: 182–203; reprinted in [Noether 1983, 271–292].

[Noether 1921a] Noether, Emmy: Über eine Arbeit des im Kriege gefallenen K. Hentzelt zur Eliminationstheorie, *Jahresbericht der Deutschen Mathematiker-Vereinigung* 30: 101; reprinted in [Noether 1983, 353].

[Noether 1921b] Noether, Emmy: Idealtheorie in Ringbereichen, *Mathematische Annalen* 83: 24-66; reprinted in [Noether 1983, 354–396].

[Noether 1923a] Noether, Emmy: Bearbeitung von Kurt Hentzelt (gefallen): Zur Theorie der Polynomideale und Resultanten, *Mathematische Annalen* 88: 53–79; reprinted in [Noether 1983, 409–435].

[Noether 1923c] Noether, Emmy: Eliminationstheorie und allgemeine Idealtheorie, *Mathematische Annalen* 90: 229-261; reprinted in [Noether 1983, 444–476].

[Noether 1924] Noether, Emmy: Abstrakter Aufbau der Idealtheorie im algebraischen Zahlkörper, *Jahresbericht der Deutschen Mathematiker-Vereinigung* 33, Abt. 2: 102; reprinted in [Noether 1983, 482].

[Noether 1925b] Noether, Emmy: Ableitung der Elementarteilertheorie aus der Gruppentheorie, *Jahresbericht der Deutschen Mathematiker-Vereinigung* 34, Abt. 2: 104.

[Noether 1927] Noether, Emmy: Abstrakter Aufbau der Idealtheorie in algebraischen Zahl- und Funktionenkörpern, *Mathematische Annalen* 96: 26–61;
reprinted in [Noether 1983, 493–528].

[Noether 1928] Noether, Emmy: Hyperkomplexe Größen und Darstellungstheorie
in arithmetischer Auffassung, *Atti Congresso Bologna*, 2, pp. 71–73; reprinted
in [Noether 1983, 560–562].

[Noether 1929a] Noether, Emmy: Hyperkomplexe Größen und Darstellungstheorie, *Mathematische Zeitschrift* 30: 641–692; reprinted in [Noether 1983, 563–
614].

[Noether 1929c] Noether, Emmy: Idealdifferentiation und Differente, *Jahresbericht der Deutschen Mathematiker-Vereinigung* 39: 17; reprinted in
[Noether 1983, 623].

[Noether 1932a] Noether, Emmy: Normalbasis bei Körpern ohne höhere Verzwei-
gung, *Journal für die reine und angewandte Mathematik* 167: 147–152; reprinted
in [Noether 1983, 624–629].

[Noether 1932b] Noether, Emmy: Hyperkomplexe Systeme in ihren Beziehungen zur kommutativen Algebra und Zahlentheorie, *Verhandlungen des Internationalen Mathematiker Kongresses Zürich* 1, pp. 189–194; reprinted in
[Noether 1983, 636–641].

[Noether 1932c] Noether, Emmy: Rezension von *Leopold Kroneckers Werke*, Bd.
5, Kurt Hensel, Hrsg., *Jahresbericht der Deutschen Mathematiker-Vereinigung*
41(Teil 2): 27.

[Noether 1933a] Noether, Emmy: Nichtkommutative Algebren, *Mathematische
Zeitschrift* 37: 514–541; reprinted in [Noether 1983, 642–669].

[Noether 1933b] Noether, Emmy: Der Hauptgeschlechtssatz für relativ-galoissche
Zahlkörper, *Mathematische Annalen* 108: 411–419; reprinted in [Noether 1983,
670–678].

[Noether 1934] Noether, Emmy: Zerfallende verschränkte Produkte und ihre
Maximalordnungen, *Exposés mathematiques publiés à la memoire de J. Her-
brand IV, Actualités scientifiques et industielles*. 148: 5–15; reprinted in
[Noether 1983, 679–689].

[Noether 1950] Noether, Emmy: Idealdifferentiation und Differente, *Journal für
die reine und angewandte Mathematik* 188: 1–21; reprinted in [Noether 1983,
690–710].

[Noether 1983] Noether, Emmy: *Gesammelte Abhandlungen, Collected Papers*,
Nathan Jacobson, ed., Heidelberg: Springer.

[Noether/Brauer 1927] Noether, Emmy und Brauer, Richard: Über minimale
Zerfällungskörper irreduzibler Darstellungen, *Sitzungsberichte der Preußischen
Akademie der Wissenschaften*, 221–228; reprinted in [Noether 1983, 552–559].

[Noether/Schmeidler 1920] Noether, Emmy und Schmeidler, Werner: Moduln in nichtkommutativen Bereichen, insbesondere aus Differential- und Differenzenausdrücken, *Mathematische Zeitschrift* 8: 1–35; reprinted in [Noether 1983, 318–352].

[M. Noether 1914] Noether, Max: Paul Gordan, *Mathematische Annalen* 75: 1–14.

[Olesko 1991] Olesko, Kathryn M.: *Physics as a calling: Discipline and practice in the Königsberg Seminar for physics*, Ithaca and London : Cornell Univ. Press.

[Osgood 1892] Osgood, William F.: The Symbolic Notation of Aronhold and Clebsch, *American Journal of Mathematics* 14(3): 251–261.

[Ostrowski 1918] Ostrowski, Alexander: Über einige Lösungen der Funktionalgleichung $\phi(x) \cdot \phi(y) = \phi(xy)$, *Acta Mathematica* 41: 271–284.

[Parikh 1991] Parikh, Carol: *The Unreal Life of Oscar Zariski*, Boston: Academic Press.

[Parshall 2015] Parshall, Karen H.: Training Women in Mathematical Research: The First Fifty Years of Bryn Mawr College (1885– 1935), *Mathematical Intelligencer* 37(2): 71–83.

[Parshall/Rice 2002] Parshall, Karen H. and Rice, Adrian, eds.: *Mathematics Unbound: The Evolution of an International Mathematical Research Community, 1800-1945*, Providence, RI: American Mathematical Society.

[Parshall/Rowe 1994] Parshall, Karen H. and Rowe, David E.: *The Emergence of the American Mathematical Research Community, 1876–1900. J.J. Sylvester, Felix Klein, and E.H. Moore*, Providence, RI: American Mathematical Society.

[Pinl/Dick 1974] Pinl, Maximilian and Dick, Auguste: Kollegen in einer dunklen Zeit, *Jahresbericht der Deutschen Mathematiker-Vereinigung* 75: 166–208.

[Pipes 2003] Pipes, Richard: *The Degaev Affair: Terror and Treason in Tsarist Russia*, New Haven, CT: Yale University Press.

[Purkert 2002] Purkert, Walter: Grundzüge der Mengenlehre – Historische Einführung, in [Hausdorff 2002, 1–90].

[Quinn et al. 1983] Quinn, Grace S., et al.: Emmy Noether in Bryn Mawr, in [Srinivasan/Sally 1983], pp. 139–146.

[Reid 1970] Reid, Constance: *Hilbert*, New York: Springer.

[Reid 1976] Reid, Constance: *Courant in Göttingen and New York: The Story of an Improbable Mathematician*, New York: Springer.

[Reid 1996] Reid, Constance: *Hilbert*, New York: Springer.

[Remmert 1995] Remmert, Volker: Zur Mathematikgeschichte in Freiburg. Alfred Loewy (1873–1935): Jähes Ende späten Glanzes, *Freiburger Universitätsblätter* 129: 81–102.

[Richardson 1934] Richardson, R.G.D.: The annual meeting in Cambridge, *Bulletin of the American Mathematical Society* 40(3): 177–188.

[Richarz 2015] Richarz, Monika: Jüdische Akademiker als Anwälte und Ärzte: Behinderte Emanzipation und berufliche Orientierung, in Christina von Braun, Hrsg., *Was war deutsches Judentum? 1870–1933*, 167–179.

[Ringer 1969] Ringer, Fritz K.: *The Decline of the German Mandarins. The German Academic Community 1890–1933*, Cambridge, MA: Harvard University Press.

[Roquette 2002] Roquette, Peter: Class Field Theory in Characteristic p, its Origin and Development; www.mathi.uni-heidelberg.de/ roquette/manu.html

[Roquette 2003] Roquette, Peter: History of Valuation Theory, Part I; online: www.mathi.uni-heidelberg.de/ roquette/manu.html

[Roquette 2004] Roquette, Peter: The Brauer-Hasse-Noether theorem in historical perspective, *Schriften der mathematisch-naturwissenschaftlichen Klasse der Heidelberger Akademie der Wissenschaften* 15: 1–92.

[Roquette 2007] Roquette, Peter: Zu Emmy Noethers Geburtstag: Einige neue Noetheriana, *Mitteilungen der DMV*, 15: 15–21.

[Roquette 2008] Roquette, Peter: Emmy Noether: Die Gutachten, (Gutachten, Begleitbriefe, Petition der Studenten, Fragebogen, Entlassungsschreiben: Geheimes Staatsarchiv Preußischer Kulturbesitz, Berlin. GStA PK, I. HA Rep. 76 Kultusministerium, Nr. 10081), https://www.mathi.uni-heidelberg.de/ roquette/gutachten/noether-gutachten.htm

[Rosenow 1998] Rosenow, Ulf: Göttinger Physik unter dem Nationalsozialismus, in *Die Universität Göttingen unter dem Nationalsozialismus*, 2te Ausgabe, H. Becker, H-J. Dahms, C. Wegeler, Hrsg., München: K.G. Saur, S. 552–587.

[Rota 2008] Rota, Gian-Carlo: *Indiscrete Thoughts*, Basel: Birkhäuser.

[Rowe 1986] Rowe, David E.: "'Jewish Mathematics' at Göttingen in the Era of Felix Klein," *Isis*, 77: 422–449.

[Rowe 1994] Rowe, David E.: Dirk Jan Struik and his Contributions to the History of Mathematics, (Introductory Essay to the Festschrift Celebrating Struik's 100th Birthday), *Historia Mathematica*, 21(3): 245–273.

[Rowe 2004] Rowe, David E.: Making Mathematics in an Oral Culture: Göttingen in the Era of Klein and Hilbert, *Science in Context*, 17(1/2): 85–129.

[Rowe 2016] Rowe, David E.: From Graz to Göttingen: Neugebauer's Early Intellectual Journey, *A Mathematician's Journeys: Otto Neugebauer and Modern Transformations of Ancient Science*, Alexander Jones, Christine Proust and John Steele, eds., Archimedes, New York: Springer, pp. 1–59.

[Rowe 2018a] Rowe, David E.: *A Richer Picture of Mathematics: The Göttingen Tradition and Beyond*, New York: Springer.

[Rowe 2018b] Rowe, David E., Hrsg..: *Otto Blumenthal, Ausgewählte Briefe und Schriften I, 1897–1918*, Mathematik im Kontext, Heidelberg: Springer.

[Rowe 2020a] Rowe, David E.: On the Pleasures and Pitfalls of Mathematical Storytelling: Conversations with Constance Reid, *Mathematical Intelligencer*, online 20 April 2020.

[Rowe 2020b] Rowe, David E.: *Emmy Noether: Mathematician Extraordinaire*, Cham: Springer Nature Switzerland.

[Rowe/Felsch 2019] Rowe, David E. und Felsch, Volkmar, Hrsg..: *Otto Blumenthal, Ausgewählte Briefe und Schriften II, 1919–1944*, Mathematik im Kontext, Heidelberg: Springer.

[Sanderson 1913] Sanderson, Mildred: Formal modular invariants with application to binary modular covariants, *Transactions of the American Mathematical Society* 14: 489–500.

[Sasaki 2002] Sasaki, Chikara: The Emergence of the Japanese Mathematical Community in the Modern Western Style, 1855–1945, [Parshall/Rice 2002, 229–252].

[Schappacher 1998] Schappacher, Norbert: Das Mathematische Institut der Universität Göttingen 1929–1950, in *Die Universität Göttingen unter dem Nationalsozialismus*, 2te Ausgabe, H. Becker, H-J. Dahms, C. Wegeler, Hrsg., München: K.G. Saur, S. 523–550.

[Schappacher 2000] Schappacher, Norbert: Das Mathematische Institut der Universität Göttingen 1929–1950 (längere Fassung).

[Schappacher 2007] Schappacher, Norbert: A Historical Sketch of B.L. van der Waerden's Work on Algebraic Geometry, 1926–1946, in [Gray/Parshall 2007, 245–283].

[Schappacher/Scholz 1992] Schappacher, Norbert und Scholz, Erhard, Hrsg.: Oswald Teichmüller – Leben und Werk, *Jahresbericht der Deutschen Mathematiker-Vereinigung* 94: 1–39.

[Schlote 1991] Schlote, Karl-Heinz: Fritz Noether–Opfer Zweier Diktaturen, *NTM. Zeitschrift für Geschichte der Naturwissenschaften, Technik und Medizin* 28: 33–41.

[Schmidt 1951] Schmidt, Erhard: Ansprachen anläßlich der Feier des 75. Geburtstages von Erhard Schmidt durch seine Fachgenossen (13.1.1951), Typoskript.

[Schmitz 2006] Schmitz, Norbert: *Moritz Abraham Stern (1807–1894). Der erste jüdische Ordinarius an einer deutschen Universität und sein populärastronomisches Werk*. Laatzen: Wehrhahn.

[Schneider 2011] Schneider, Martina: *Zwischen zwei Disziplinen. B.L. van der Waerden und die Entwicklung der Quantenmechanik*, Heidelberg: Springer.

[Schneider 2021] Schneider, Martina: Van der Waerden, Noether und die Noether-Schule, to appear in [Koreuber 2021].

[Scholz 2008] Scholz, Erhard: Hausdorffs Blick auf die entstehende algebraische Topologie, in [Hausdorff 2008, 865–892].

[Schüddekopf/Zieher 2019] Schüddekopf, Sandra und Zieher, Anita: Mathematische Spaziergänge mit Emmy Noether (Stücktext), *portraittheater Vienna*.

[Schüddekopf/Zieher 2020] Schüddekopf, Sandra und Zieher, Anita: Diving into Math with Emmy Noether (Script), *portraittheater Vienna*.

[Seidl, et al. 2018] Seidl, Ernst; Loose, Frank; Bierende, Edgar (Hrsg.): *Mathematik mit Modellen. Alexander von Brill und die Tübinger Modellsammlung*. Tübingen: Universität Tübingen.

[Segal 2003] Segal, Sanford L.: *Mathematicians under the Nazis*, Princeton: Princeton University Press.

[Severi 1933] Severi, Francesco: Über die Grundlagen der algebraischen Geometrie, *Abhandlungen aus dem mathematischen Seminar der Hambrgischen Universität* 9: 335–364.

[Shen 2019] Shen, Qinna: A Refugee Scholar from Nazi Germany: Emmy Noether and Bryn Mawr College, *The Mathematical Intelligencer* 41(3): 1–14.

[Shiryaev et al. 2000] Shiryaev, A.N., et al.: *Kolmogorov in Persepctive*, Providence, RI: American Mathematical Society.

[Siegmund-Schultze 2001] Siegmund-Schultze, Reinhard: *Rockefeller and the Internationalization of Mathematics Between the Two World Wars*, Basel: Birkhäuser.

[Siegmund-Schultze 2009] Siegmund-Schultze, Reinhard: *Mathematicians Fleeing from Nazi Germany: Individual Fates and Global Impact*, Princeton: Princeton University Press.

[Siegmund-Schultze 2011a] Siegmund-Schultze, Reinhard: Göttinger Feldgraue, Einstein und die verzögerte Wahrnehmung von Emmy Noethers Sätzen über invariante Variationsprobleme (1918), *Mitteilungen der DMV*, 19: 100–104.

[Siegmund-Schultze 2011b] Siegmund-Schultze, Reinhard: Bartel Leendert van der Waerden (1903–1996) im Dritten Reich: Moderne Algebra im Dienst des Anti-Modernismus? *Fremde Wissenschaftler im Dritten Reich: die Debye-Affäre im Kontext*, Dieter Hoffmann, Mark Walker, Hrsg., Göttingen: Wallstein, pp. 200–229.

[Siegmund-Schultze 2015] Siegmund-Schultze, Reinhard: Van der Waerden in the Third Reich, *Notices of the AMS*, 62(8): 924–929.

[Siegmund-Schultze 2016] Siegmund-Schultze, Reinhard: Mathematics Knows No Races: a political speech that David Hilbert planned for the ICM in Bologna in 1928, *Mathematical Intelligencer*, 38(1): 56–66.

[Soifer 2015] Soifer, Alexander: *The Scholar and the State: In Search of Van der Waerden*, New York. Springer.

[Srinivasan/Sally 1983] Srinivasan, Bhama and Judith D. Sally, eds., *Emmy Noether in Bryn Mawr. Proceedings of a Symposium Sponsored by the Association for Women in Mathematics in Honor of Emmy Noether's 100th Birthday*, New York: Springer.

[Steinitz 1910] Steinitz, Ernst: Algebraische Theorie der Körper, *Journal für die reine und angewandte Mathematik* 137: 167–309.

[Steinitz 1930] Steinitz, Ernst: *Algebraische Theorie der Körper*, Mit einem Anhang: Abriss der Galoisschen Theorie von Reinhold Baer und Helmut Hasse, Berlin: De Gruyter.

[Takagi 1920] Takagi, Teiji: Über eine Theorie des relativ abelschen Zahlkörpers, *Journal of the College of Science, Imperial University of Tokyo* 41: 1–133.

[Takagi 1922] Takagi, Teiji: Über das Reziprozitätsgesetz in einem beliebigen algebraischen Zahlkörper, *Journal of the College of Science, Imperial University of Tokyo* 44: 1–50.

[Takagi 1935] Takagi, Teiji: Reminiscences and Perspectives, in *Miscellaneous Notes on Mathematics*, Tokyo.

[Taussky 1981] Taussky, Olga: My Personal Recollections of Emmy Noether, in [Brewer/Smith 1981, 79–92].

[Taussky-Todd 1985] Taussky-Todd, Olga: An Autobiographical Essay, in *Mathematical People: Profiles and Interviews*, D.J. Albers and G.L. Alexanderson, eds., Boston: Birkhäuser, pp. 309–336.

[Tobies 2003] Tobies, Renate: Briefe Emmy Noethers an P. S. Alexandroff, *NTM. Zeitschrift für Geschichte der Naturwissenschaften, Technik und Medizin* 11: 100–115.

[Tobies 2019] Tobies, Renate: *Felix Klein. Visionen für Mathematik, Anwendungen und Unterricht*, Heidelberg: Springer.

[Tollmien 1990] Tollmien, Cordula: „Sind wir doch der Meinung, dass ein weiblicher Kopf nur ganz ausnahmsweise in der Mathematik schöpferisch tätig sein kann. . . ,"Emmy Noether 1882–1935, *Göttinger Jahrbuch*, 38: 153–219.

[Tollmien 2016a] Tollmien, Cordula: „Das mathematische Pensum hat sie sich durch Privatunterricht angeeignet"– Emmy Noethers zielstrebiger Weg an die Universität, in Andrea Blunck, Renate Motzer, Nicola Oswald, Hrsg., *Tagungsband zur Doppeltagung: Frauen in der Mathematikgeschichte & Herbsttreffen Arbeitskreis Frauen und Mathematik*, Hildesheim: Franzbecker, S. 1–12.

[Tollmien 2016b] Tollmien, Cordula: „Die Weiblichkeit war nur durch Fräulein Emmy Noether vertreten"– Die Mathematikerin Emmy Noether, *Göttinger Stadtgespräche*, Christiane Freudenstein, Hrsg., Göttingen: Vandenhoeck & Ruprecht, S. 185–193.

[Tsen 1933] Tsen, Chiungtze C.: Divisionsalgebren über Funktionenkörpern, *Nachrichten der Königlichen Gesellschaft der Wissenschaften zu Göttingen*, Mathematisch-Physikalische Klasse, 335–339.

[van Dalen 2013] van Dalen, Dirk: *L.E.J. Brouwer–Topologist, Intuitionist, Philosopher. How Mathematics is Rooted in Life*, London: Springer.

[van der Waerden 1927] van der Waerden, B.L.: Zur Nullstellentheorie der Polynomideale, *Mathematische Annalen* 96: 183–208.

[van der Waerden 1930a] van der Waerden, B.L.: Kombinatorische Topologie, 39: *Jahresbericht der Deutschen Mathematiker-Vereinigung* 121–139.

[van der Waerden 1930/31] van der Waerden, B.L.: *Moderne Algebra*, 2 Bde., Berlin: Julius Springer.

[van der Waerden 1932] van der Waerden, B.L.: *Die Gruppentheoretische Methode in der Quantenmechanik*, Berlin: Julius Springer.

[van der Waerden 1935] van der Waerden, B.L.: Nachruf auf Emmy Noether, *Mathematische Annalen* 111: 469–476.

[van der Waerden 1971] van der Waerden, B.L.: The Foundation of Algebraic Geometry from Severi to André Weil, *Archive for History of Exact Sciences* 7(3): 171–180.

[van der Waerden 1975] van der Waerden, B.L.: On the Sources of my Book *Moderne Algebra*, *Historia Mathematica* 2: 31–40.

[van der Waerden 1997] van der Waerden, B.L.: Meine Göttinger Lehrjahre, *Mitteilungen der DMV* 5(2): 20–27.

[Weil 1992] Weil, André: *The Apprenticeship of a Mathematician*, Basel: Birkhäuser.

[Weyl 1927] Weyl, Hermann: Quantenmechanik und Gruppentheorie, *Zeitschrift für Physik* 46: 1–46; reprinted in [Weyl 1968, 3: 90–135].

[Weyl 1934/35] Weyl, Hermann: *The Structure and Representation of Continuous Groups*, Lithographed notes by R. Brauer of Weyl's lectures at the Institute of Advanced Study, Princeton.

[Weyl 1935] Weyl, Hermann: Emmy Noether, (Memorial Address, 26 April 1935), *Scripta Mathematica*, 3: 201–220; reprinted in [Weyl 1968, 3: 425–444].

[Weyl 1944] Weyl, Hermann: David Hilbert and his Mathematical Work, *Bulletin of the American Mathematical Society*, 50(9): 612–654; reprinted in [Weyl 1968, 4: 130–172].

[Weyl 1968] Weyl, Hermann: *Gesammelte Abhandlungen*, 4 vols., ed. K. Chandrasekharan. Berlin: Springer-Verlag.

[Wilhelm 1979] Wilhelm, Peter: *Die Synogogengemeinde Göttlingen, Rosdorf und Geismar, 1850–1942*, Göttingen: Vandenhoeck & Ruprecht.

[Yandell 2002] Yandell, Ben H.: *The Honors Class. Hilbert's Problems and their Solvers*, Natick, Mass.: AK Peters.

Name Index

Printed in the United States
by Baker & Taylor Publisher Services